Inventory of Biodiversity Today

Series Editor
Marie-Christine Maurel

Inventory of Biodiversity Today

New Methods and Discoveries

Edited by

Violaine Nicolas

iSTE

WILEY

First published 2024 in Great Britain and the United States by ISTE Ltd and John Wiley & Sons, Inc.

ISTE Ltd
27-37 St George's Road
London SW19 4EU
UK

www.iste.co.uk

John Wiley & Sons, Inc.
111 River Street
Hoboken, NJ 07030
USA

www.wiley.com

Library of Congress Control Number: 2024938962

British Library Cataloguing-in-Publication Data
A CIP record for this book is available from the British Library
ISBN 978-1-78630-958-7

Contents

Introduction

Inventory of Biodiversity Today: New Methods and Discoveries

Today, as we face the prospect of the sixth mass extinction, the term "biodiversity" has gained popularity, and is widely used. Coined in the 1980s (Wilson 1988), the word is a contraction of the phrase "biological diversity". At the Earth Summit in Rio de Janeiro in 1992, its definition was set in stone:

> '*Biological diversity*' means the variability among living organisms from all sources including, *inter alia*, terrestrial, marine and other aquatic ecosystems and the ecological complexes of which they are part; this includes diversity within species, between species and of ecosystems.

Today, we know of only 2.2 million species (Roskov et al. 2014), which is around 20% of the 11 million species estimated to exist. Certain studies – notably those which take account of microbial diversity – place the number far higher, proposing that there are somewhere between one and six billion species (Larsen et al. 2017). In addition, often, we have very limited knowledge of these species. With respect to over half the known macro-organisms, we have only brief descriptions found in dated publications; with respect to unculturable microbes, we are only aware of a few molecules (Troudet et al. 2017).

Introduction by Violaine NICOLAS.

At the present juncture, it is absolutely crucial that we step up the work of inventorying our world's biodiversity, with a view to describing the remaining 80% +/–, not only to assuage our own thirst for knowledge and understanding of the world around us and how it evolves, but also because such knowledge could have very concrete implications in our day-to-day lives. For example, recent work has shown that previously unknown, or poorly known, organisms may have critical importance for our societies. The recent Covid-19 pandemic, which brought the world to a standstill for months on end, is an unfortunate example. It shed light on how little we understand the wide variety of coronaviruses and their hosts (Hassanin et al. 2020). Issues such as this may affect not only human health, but also the health of the plants we eat. Think for example, of the plant viruses and viroids about which little is known, but which have been seen to devastate certain crops (Maurel 2018). In addition, it has recently been shown that certain poorly understood micro-organisms have a major role to play in the biogeochemical cycles which govern our world, such as the carbon cycle, the methane cycle, the nitrogen cycle or the sulfur cycle (Lannes et al. 2021). Biodiversity is crucial for us in our daily lives: the oxygen we breathe; the food and water we consume; the drugs which treat our ills; the raw materials we use to clothe ourselves (cotton, hemp, wool, etc.) or build shelters (wood); crop pollination; soil fertilization; climate regulation; soil formation and prevention of soil erosion; the natural purification of water by plants in humid environments; the prevention of water rises and floods by wetlands. All of this is gifted by nature (Grandcolas and Marc 2023). Plainly then, describing and understanding this biodiversity is crucial for our very survival.

How, though, are we to approach this gargantuan task of describing the myriad species, while the number of taxonomists – scientists specializing in this domain – is constantly decreasing? Each year, between 15,000 and 20,000 new species are described by between 30,000 and 40,000 taxonomists, both professional and amateurs; this equates to around one new species every 30 minutes (Miralles et al. 2020). Thus, at the current rate, it would take at least several centuries to fully accomplish the task of completing all taxonomic descriptions (Costello et al. 2013). However, looking down the barrel of the sixth mass extinction event, we realize that time is a luxury we do not have. Numerous species – the majority of species, even – are likely to die out before they can be described, similarly to what are known as the "dark extinctions" of the pre-taxonomic period, before the 1800s (Boehm and Cronk 2021).

The aim of this book is to demonstrate how today's scientists go about dealing with this immense task. In particular, the approach involves organizing large-scale exploration and inventorying campaigns, which may take place in locations and environments that are nearby and easily accessible, or that are distant or difficult to access such as the ocean depths or the poles. Another strategy is to set up participative science projects, in which ordinary citizens are invited to contribute to the data collection. There are a wide variety of methods used to collect and analyze samples, and numerous forms of data associated with specimens are now systematically recorded (e.g. temperature, pH, conductivity, salinity and turbidity of the water, current direction and velocity, partial pressure of dissolved gases), so as to better understand the interactions between the organisms and their environment. In addition, new methods of data acquisition and processing have been developed, enabling scientists to acquire increasingly precise and unprecedented morphological or genetic data, on an ever-growing number of specimens. For instance, computed microtomography, using X-rays, allows us to view the internal microstructures and create 3D models of these anatomic structures, on both current and fossilized specimens, without causing physical damage to the specimens. Observation with ultraviolet light and environmental scanning electron microscopy (ESEM) also give us access to morphological details without harming the specimens. Approaches based on genetic data have also flourished, and today, they can be used to acquire essential biodiversity data not just from specimens preserved, for varying periods of time, in the collections of natural history museums (barcoding, museum genomics), but also directly from environmental samples such as water or soil samples (metabarcoding).

The development of all of these techniques means that, today, it is possible to generate a vast quantity of data on specimens, which must then be interwoven to define and describe the various species, by means of integrative taxonomy.

Thanks to scientists' activity and recent technical advances, today we are producing unprecedented quantities of data, which need to be better handled and exploited, taking account of potential sources of skew. In order to better exploit these data sources, it is essential to continue to develop open science, which is a means of conducting research combining production and widespread dissemination, with the goal being to enable as many people as possible to freely access, reuse, build and distribute these resources, unencumbered. Not only does such an approach allow scientists in varying disciplines to pool their knowledge in order to answer complex questions, it also popularizes the research results, so that all of society can have access to them, make use of them, and contribute to the process, be it during the acquisition or data analysis phase.

References

Boehm, M.M.A. and Cronk, Q.C.B. (2021). Dark extinction: The problem of unknown historical extinctions. *Biol. Lett.*, 17(3), 20210007.

Costello, M.J., May, R.M., Stork, N.E. (2013). Can we name Earth's species before they go extinct? *Science*, 339(6118), 413–416.

Grandcolas, P. and Marc, C. (2023). *Tout comprendre (ou presque) sur la biodiversité*. CNRS Éditions, Paris.

Hassanin, A., Grandcolas, P., Veron, G. (2020). Covid-19: Natural or anthropic origin? *Mammalia*, 85(1), 1–7.

Lannes, R., Cavaud, L., Lopez, P., Bapteste, E. (2021). Marine ultrasmall prokaryotes likely affect the cycling of carbon, methane, nitrogen, and sulfur. *Genome Biol. Evol.*, 13(1), evaa261.

Larsen, B.B., Miller, E., Rhodes, M.K., Wiens, J.J. (2017). Inordinate fondness multiplied and redistributed: The number of species on earth and the new pie of life. *Q. Rev. Biol.*, 92(3), 229–265.

Maurel, M.-C. (2018). À la frontière du vivant : les viroïdes. *The Conversation* [Online]. Available at: https://theconversation.com/a-la-frontiere-du-vivantles-viro-des-90500.

Miralles, A., Bruy, T., Wolcott, K., Scherz, M.D., Begerow, D., Beszteri, B., Bonkowski, M., Felden, J., Gemeinholzer, B., Glaw, F. et al. (2020). Repositories for taxonomic data: Where we are and what is missing? *Syst. Biol.*, 69(6), 1231–1253.

Roskov, Y., Kunze, T., Orrell, T., Abucay, L., Paglinawan, L., Culham, A., Bailly, N., Kirk, P.M., Bourgoin, T., Baillargeon, G. et al. (2014). *Species 2000 & ITIS Catalogue of Life, 2014 Annual Checklist*. Naturalis, Leiden.

Troudet, J., Grandcolas, P., Blin, A., Vignes-Lebbe, R., Legendre, F. (2017). Taxonomic bias in biodiversity data and societal preferences. *Sci. Rep.*, 7(1), 9132.

Wilson, E.O. (1988). *Biodiversity*. National Academy Press, Washington.

1

Scientific Exploration Campaigns to Inventory Existing Biodiversity and Hasten the Discovery of New Species

1.1. Introduction

Life on Earth first emerged at least 3.5 billion years ago. Since then, it has been constantly evolving, which has led to the phenomena of both diversification and extinction. The fossil records of marine and terrestrial organisms show that the diversity of life, both in the seas and on land, increased exponentially after the end of the Precambrian (Benton 1995), 540 million years ago. Such diversification, though, was punctuated by mass extinctions; the term "biotic crisis" is also employed. Most such extinction events struck marine and terrestrial organisms simultaneously. Within the geological time scale (Bergström et al. 2009), we see no fewer than five major crises, occurring during the Ordovician, the Devonian, the Permian, the Triassic–Jurassic transition and the Cretaceous–Tertiary transition. For example, the Permian–Triassic extinction, which took place some 252 million years ago, saw 95% of marine species and 70% of terrestrial species die out. During the Cretaceous–Tertiary extinction 66 million years ago, during which the dinosaurs rapidly perished, the mammals – who, up until that point, had coexisted with the dinosaurs – suddenly had the place to themselves (so to speak), and we see the group massively diversify. It is plain that the variety of living organisms on planet Earth has resulted from a lengthy and gradual, dynamic process, punctuated by crises. For over a

---

Chapter written by Line LE GALL.

quarter of a century, researchers have been asking whether the particularly proliferous demographic expansion of humans during the Holocene has profoundly altered Earth itself (Myers 1990), leading to a sixth mass extinction event. The extinction rates calculated by biologists tend to demonstrate that this crisis will occur at an even faster rate than the previous five biotic crises (Barnosky et al. 2011), erasing the planet's biological heritage even before it has been fully described.

1.2. A historical overview of diversity inventory

Reindeer, aurochs, mammoths, horses, bison… as early as 18,000 years ago, our prehistoric ancestors made paintings of animals. Whilst we do not yet know, in detail, what these paintings signified, it is undeniable that they constitute the earliest form of inventory of the surrounding biological diversity. Even in the earliest known literary work, the Epic of Gilgamesh (18th century BCE), the landscape in which the action takes place is described using imagery of fauna (for example, the buffalo) and flora (for example, the cedar). In Ancient times, Aristotle (384–322 BCE), in the Organon, devotes a great deal of time to the observation of animated beings (animals), drawing a contrast with inanimate beings (plants). Aristotle's pupil, Theophrastus (372–288 BCE), followed in Aristotle's footsteps, and in particular, developed the discipline of botany, producing a reasoned inventory of the plant life found in the Mediterranean Basin. These two authors offer us detailed observations and descriptions of various animal and plant species; they began to document and classify plants and animals. From the 16th century onward, major explorations by sea led to the discovery of new lands and previously unknown species. By way of example, Captain James Cook's voyages in the 18th century greatly enriched our knowledge of biodiversity; Cook brought back numerous specimens, which needed to be named. Around the same time, a method emerged for scientific description of the living world, gradually coalescing into classification and nomenclature systems with which to organize known species. The seminal work of Carl Linnaeus (also known as Carl von Linné), *Systema Naturae*, published in 1735, laid the foundations for modern taxonomy and greatly simplified the way in which species were named, proposing the binomial system and comprising a genus name, followed by a species label. This did away with the lengthy descriptive names in use until that point. The most active period for species description was the 19th century; the natural history collections of the major museums are extremely rich in specimens from that period, which was marked by large-scale scientific expeditions. Charles Darwin's observations during his voyage aboard the HMS Beagle are famed, in view of the major advances they brought about in the understanding of the evolution and diversity of species. This work led to the publication of Darwin's *On the Origin of Species* in 1859, advancing a convincing scientific theory on the origin of the diversity of the living world. France also played an important part: numerous explorations greatly enriched the collections of the National Museum of Natural History. These included Napoleon Bonapart's Egyptian Campaign (1798–1801), which included academics and artists who were charged with studying and documenting the history, architecture, fauna, flora and culture of Ancient Egypt. The expedition of *La Coquille* (1822–1825) is another example. Under the command of Louis-Isidore Duperrey, *La Coquille* sailed the seas of the world, notably focusing on the Pacific Ocean and the islands of the South Pacific. Naturalist research was carried out jointly by Jules Dumont d'Urville, who was in charge of botany and entomology, and René Primevère Lesson, who was in charge of zoology. These studies led to the discovery of numerous new species. After the tropics, its was the poles which captured explorers' attention, around 1900. In France, Commander Jean-Baptiste Charcot set out to explore both the South and the North Pole. The two World Wars hindered scientific exploration, but also led to major technological breakthroughs, such as the development of electron microscopy. Thereafter, biologists have largely turned their attention to the study of biochemistry and cell biology, to the detriment of species inventory.

1.3. The advent of molecular taxonomy

The discovery of DNA as the carrier of genetic information, and then the discovery of its structure, but above all, the invention of DNA sequencing techniques in the 1970s (Sanger et al. 1977), were giant leaps forward in molecular biology and genomics. DNA sequencing greatly changed the way in which we understood biodiversity, revealing entire sectors of the living world, thanks to phylogenetic methods, which can compare similar DNA sequences with a view to inferring the evolutionary links between specimens. These methods soon led to the discovery of a third domain of the living world: Archaea (Woese and Fox 1977), which, alongside true bacteria, make up the group of procaryotes (i.e. cells without a nucleus), in contrast to eucaryotes (cells which do have a nucleus). Aside from the discovery of this new sector of the living world, the use of DNA sequences as markers to differentiate species considerably sped up the scientific inventory of biodiversity, notably revealing a number of new cryptic species (Blaxter 2004). Paul Hebert (Hebert et al. 2003) proposed using DNA sequences in the same way as a barcode is used in the commercial world, to identify organisms, and thus create a library of the living (Le Gall et al. 2017). Utilizing technological advances, which allow for the handling of massive databases of sequences, the "Earth BioGenome" project (Lewin et al. 2022) aims to create a library, not for short genetic sequences, but for complete reference genomes (Earth BioGenome Project n.d.). In France, the ATLASea project will contribute to the broader international project, by sequencing the reference genome for 4,500 marine organisms.

1.4. Biodiversity: the emergence of a concept in the face of the crisis

The term "biodiversity", which is now widely used, is a very recent concept, which emerged some thirty years ago, when we became aware of the looming threat of an ecological crisis, caused by human activities. Raymond F. Dasmann is credited with coining the term "biological diversity" (Dasmann 1970), which was then taken up by Thomas Lovejoy, the then-Director of the World Wildlife Fund-US conservation program, who used the term in two 1980 publications. However, it was the Convention on Biological Diversity, ratified at the Earth Summit in Rio de Janeiro in 1992, that popularized the term "biological diversity", defining it as follows:

> '*Biological diversity*' means the variability among living organisms from all sources including, *inter alia*, terrestrial, marine and other aquatic ecosystems and the ecological complexes of which they are part; this includes diversity within species, between species and of ecosystems.

The contracted form of "biological diversity" as "biodiversity" then gained widespread use. It is also a concept which has been enriched over time and now, beyond referring to the diversity of the living world, it also includes the way in which we perceive that biodiversity, and in particular, the functions and uses we make of it, not forgetting the symbolic aspects attaching to it.

The Convention on Biological Diversity also established a framework for research and training (article 12), which stipulates that each signatory country shall: “Establish and maintain programmes [sic] for scientific and technical education and training in measures for the identification, conservation and sustainable use of biological diversity”. Though numerous programs and initiatives the world over are devoted to the task, the inventory of biodiversity is far from complete.

1.5. An incomplete inventory of diversity

Inventorying biodiversity consists of identifying and cataloging the species present in a given ecosystem. From the highest peaks on the planet to the deepest trenches of the oceans, to-date, scientists have recorded some two million species (ITIS SPECIES 2000, 2019; Annual Checklist: https://www.catalogueoflife.org/) inhabiting Earth. Of these, 1.3 million are animals, 375,000 are plants (Christenhusz and Byng 2016), 135,000 are fungi, 10,000 are bacteria, 3,000 are viruses, and the remaining 27,000 are chiefly single-celled organisms, often grouped together under the umbrella term *Protist*, which inadequately reflects their broad divergence, as they account for some ten evolutionary lineages (Adl et al. 2019); they are often as different from one another as plants are from animals.

On average, 18,000 new species are described each year (around 2,300 of these being marine species (Bouchet et al. 2023)), so plainly, the inventory of existing biodiversity is nowhere near complete. Whilst it is a time-consuming task to catalog the species which have been formally described to-date, estimating the number of species yet to be described is a risky business. In the 1980s, Terry Erwin (1982) designed an experiment to estimate the diversity of insects. Fumigating the tree canopy in a forest in Panama, he counted the number of insects per species of tree. Then, using the data on the specificity of insects with respect to the trees that they inhabit, he calculated that this group could include up to 30 million species, as opposed to 1.5 million – the generally accepted figure at the time. Since that evaluation, numerous estimations of biodiversity have been carried out, leading to figures which vary by more than an entire order of magnitude. Mora et al. estimate biodiversity at 8.7 ± 1.2 million species, while Larsen et al. estimate the number as being several billion species, the vast majority (70 to 90%) being bacteria.

In parallel to estimations of the number of species present on Earth, there is also the matter of estimating the number of species under threat. The IPBES Report published in 2019 puts forward the worrying figure of one million species under threat of extinction (Tollefson 2019). A recent study seems to suggest that species which have not yet been officially described are even more threatened than those which have (Liu et al. 2022).

In this context, we can clearly understand the urgent need to continue investigating biodiversity, with a view to stepping up the description of new species and swiftly establishing conservation measures to protect that diversity.

1.6. The place of scientific exploration campaigns, run by the French Natural History Museum, in the inventory of biodiversity

Following in a long tradition of naturalist expeditions, the French Natural History Museum runs numerous ambitious scientific exploration programs, all over the world. There is a particular focus on so-called biodiversity hotspots, which are located mainly in the tropical regions (Tittensor et al. 2010). For example, we could point to the *Tropical Deep-Sea Benthos* program, launched in 1976 with the IRD (Institute for Research and Development), which explores and analyses biodiversity in the deep waters of the intertropical zone. For over 45 years, thousands of operations have been carried out as part of this program, in the Pacific and the Caribbean. The program "*La Planète revisitée*" (The Planet Revisited), launched by the expedition to Espiritu Santo in 2006, aims to inventory the natural heritage in regions where it is astonishingly rich, exploring both terrestrial and marine biodiversity. Thus, taxonomists have studied forests and coasts from New Caledonia to Guyana, Corsica, Papua New Guinea and Madagascar, with a fine-toothed comb. These programs enrich the collections and focus primarily on taxa which have received little attention because they are often not highly striking; these are known as neglected taxa. Among the numerous targets are lichens, algae, fungi, insects, mollusks, crustaceans, annelids, and many other groups which are even less well known. The specimens harvested continue to enrich collections and scientific research the world over. Since 2010, over 20% of new marine species described in the world have been discovered through the Museum's campaigns.

1.7. Innovations to speed up the description of species

Exploration of biodiversity is an activity carried out using a methodology and an approach which, all told, have changed little in over two centuries. Even today, it is the variety of techniques for trapping and collecting samples which ensures the success of a campaign! However, numerous recent innovations have helped improve our ability to deploy sampling methods over an extended period of time and an extended space. The sites to be explored are chosen on the basis of increasingly accurate maps, thanks to satellite teledetection techniques, which are capable of acquiring data on a massive scale and in environments which are difficult to access. However, whilst these methods are particularly effective for submerged lands, they are far less effective for oceanic environments, in which bathymetric data are mainly acquired by the use of multibeam sonar, requiring oceanographic vessels to be deployed to acquire the data. In addition, the use of online databases and data-sharing platforms facilitates collaboration among scientists all around the world. This can help to identify the most relevant areas to explore, either because they have had few or no investigations carried out thus far, or else because they have been found to host extraordinary biodiversity.

Once the samples have been collected, taxonomists identify them by studying a range of characteristics. The data collected, including information on habitat, geographic distribution and ecological interactions, are recorded in databases which should, ideally, conform to the FAIR principle (which refers to data that are Findable, Accessible, Interoperable and Reusable). Artificial intelligence (AI) is also being increasingly used to analyze vast quantities of data (the term "big data" is used, as is "high-throughput data"). For example, machine-learning algorithms can be used to rapidly classify images of species, which speeds up the process of identifying organisms in field studies.

Lay citizens can also contribute to the collection of data on biodiversity, which are often beyond the reach of researchers alone. These participatory science projects set up new means of circulation of knowledge between scientists and society at large. They often involve the use of mobile apps to report observations of species, meaning data can be acquired on a large scale as to the distribution of species, and changes over time can be monitored. However, in order to bring about a significant advance in the description of biodiversity, they must, beyond simple observation, allow for sampling of material which can serve as the basis for the descriptions imposed by the nomenclature codes governing the description of new species. For instance, to supplement operations as part of the "*La Planète revisitée*" project in New Caledonia, a participatory project known as the "*quinzaine des nudibranches*" was organized to draw up an inventory of sea slugs. Divers are fascinated by these organisms because of their beauty, original shapes and vibrant colors. In total, 28 local naturalists actively participated in this quest, requiring good training to locate the individuals, as the animals are timid by nature. The involvement of these aware volunteers represented the equivalent of 263 people per day across 50 sites on the Nouméa Peninsula, the lagoon and the reef barrier. With the support of seven researchers, one of whom is a renowned specialist on sea slugs, their work helped to document 352 species, around 80 of which were new, a priori. This experiment in participatory inventory proved to be highly fruitful, and it would be worth repeating in other territories or other seasons.

1.8. Challenges and issues surrounding scientific exploration campaigns

The organizing of scientific explorations involves overcoming still more challenges. Besides the financial costs and logistical requirements which have always been in place, it is now also crucial to take account of the ethical dimension – notably in relation to the protocols for collection, the use of animals for research purposes, and the rights of local communities. The Nagoya protocol, adopted in 2010, and implemented in 2014, regulates access and benefit-sharing (ABS) stemming from the use of genetic resources and associated traditional knowledge. It structures the relations between suppliers and users of genetic resources and the associated traditional knowledge. In practice, each mission is prepared in consultation with the local actors to construct a shared project in which all parties agree on the objectives, means of action and the way in which the elements collected are to be distributed. The collections and knowledge gleaned from expeditions are made public and accessible to all, in accordance with the open-science approach. Thus, it is crucial to ensure conformity with the regulations at every step of the project – in particular, during collection. Such collection must be respectful of the environment, limiting its environmental impact to the fullest extent possible.

1.9. Conclusion

Scientific exploration campaigns are essential to the work of inventorying the planet's biodiversity, and encouraging the discovery of new species. However, there is still a long way to go before that inventory will be sufficiently complete to be robust. This lack of completeness is a hindrance to scientists, who may not have a realistic view of the web of interactions amongst species, as they can only look at 10% of those species; it also represents a stumbling block for biologists working in conservation. The IUCN, an international organization which sets a conservation status for threatened or endangered species, only does so for species which have been formally described. Thus, species which have not been described are precluded from conservation measures. It is only by continuing our efforts and continuing to organize scientific exploration campaigns that we can continue to acquire valuable knowledge about our biodiversity, with a view to preserving it, and so, protecting our natural heritage for future generations.

1.10. References

Adl, S.M., Bass, D., Lane, C.E., Lukeš, J., Schoch, C.L., Smirnov, A., Agatha, S., Berney, C., Brown, M.W., Burki, F. et al. (2019), Revisions to the classification, nomenclature, and diversity of eukaryotes. *J. Eukaryot. Microbiol.*, 66, 4–119. doi: 10.1111/jeu.12691.

Barnosky, A.D., Matzke, N., Tomiya, S., Wogan, G.O.U., Swartz, B., Quental, T.B., Marshall, C., McGuire, J.L., Lindsey, E.L., Maguire, K.C. et al. (2011). Has the Earth's sixth mass extinction already arrived? *Nature*, 471(7336), 51–57.

Benton, M.J. (1995). Diversification and extinction in the history of life. *Science*, 268(5207), 52–58.

Bergström, S.M., Chen, X., Gutiérrez-Marco, J.C., Dronov, A. (2009). The new chronostratigraphic classification of the Ordovician System and its relations to major regional series and stages and to $\delta^{13}C$ chemostratigraphy. *Lethaia*, 42(1), 97–107.

Blaxter, M.L. (2004). The promise of a DNA taxonomy. *Philos. Trans. R. Soc. Lond. B. Biol. Sci.*, 359(1444), 669–679.

Bouchet, P., Decock, W., Lonneville, B., Vanhoorne, B., Vandepitte, L. (2023). Marine biodiversity discovery: The metrics of new species descriptions. *Frontiers in Marine Science*, 10, 929989.

Christenhusz, M.J.M. and Byng, J.W. (2016). The number of known plants species in the world and its annual increase. *Phytotaxa*, 261(3), 201–217.

Dasmann, R.F. (1970). *A Different Kind of Country*. Collier Books/Collier-Macmillan, New York/London.

Earth BioGenome Project (n.d.). [Online]. Available at: https://www.earthbiogenome.org/.

Erwin, T.L. (1982). Tropical forests: Their richness in Coleoptera and other arthropod species. *The Coleopterists Bulletin*, 36(1), 74–75.

Hebert, P.D., Cywinska, A., Ball, S.L., deWaard, J.R (2003). Biological identifications through DNA barcodes. *Proc. Biol. Sci.*, 270(1512), 313–321.

ITIS SPECIES 2000 (2019). Catalogue of Life: 2019 Annual Checklist [Online]. Available at: http://www.catalogueoflife.org/annual-checklist/2019/info/totals.

Larsen, B.B., Miller, E.C., Rhodes, M.K., Wiens, J.J. (2017). Inordinate fondness multiplied and redistributed: The number of species on Earth and the new pie of life. *The Quarterly Review of Biology*, 92(3), 229–265. doi: 10.1086/693564.

Le Gall, L., Delsuc, F., Hourdez, S., Lecointre, G., Rasplus, J.-Y. (2017). Toward the DNA Library of Life. *European Journal of Taxonomy*, 266, 1–9.

Lewin H.A., Richards, S., Lieberman Aiden, E., Allende, M.L., Archibald, J.M., Bálint, M., Barker, K.B., Baumgartner, B., Belov, K., Bertorelle, G. et al. (2022). The Earth BioGenome Project 2020: Starting the clock. *Proceedings of the National Academy of Sciences*, 119(4), e2115635118.

Liu, J., Slik, F., Zheng, S., Lindenmayer, D.B. (2022). Undescribed species have higher extinction risk than known species. *Conservation Letters*, 15(3), e12876.

Mora, C., Tittensor, D.P., Adl, S., Simpson, A.G.B., Worm, B. (2011). How many species are there on Earth and in the ocean? *PLoS Biology*, 9(8), e1001127. doi: 10.1371/journal.pbio.1001127.

Myers, N. (1990). Mass extinctions: What can the past tell us about the present and the future? *Palaeogeography, Palaeoclimatology, Palaeoecology*, 82(1/2), 175–185.

Sanger, F., Nicklen, S., Coulson, A.R. (1977). DNA sequencing with chain-terminating inhibitors. *Proceedings of the National Academy of Sciences*, 74(12), 5463–5467.

Tittensor, D.P., Mora, C., Jetz, W., Lotze, H.K., Ricard, D., Berghe, E.V., Worm, B. (2010). Global patterns and predictors of marine biodiversity across taxa. *Nature*, 466, 1098–1101.

Tollefson, J. (2019). Humans are driving one million species to extinction. *Nature*, 569(7755), 171–171.

Woese, C.R. and Fox, G.E. (1977). Phylogenetic structure of the prokaryotic domain: The primary kingdoms. *Proceedings of the National Academy of Sciences*, 74(11), 5088–5090.

2

Half a Century of Naturalist Exploration of Upper Bathyal Benthic Environments: Ruptures and Continuities

2.1. The deep ocean: we have barely begun to plumb the depths

As can be seen from the current state of political affairs (French Government press releases dated 8 June 2023, and the Secretariat General for the Ocean, dated 21 January 2023), our knowledge of the deep ocean beds, defined as those zones of the ocean where light no longer penetrates to allow photosynthesis, is still highly fragmentary. The question of ocean cartography illustrates this lack of knowledge. The ocean, which covers over 70% of the planet's surface, has a highly irregular bed. The average depth is around 3,800 meters (as compared to an average altitude of only 850 meters in the case of lands above sea level). However, the estimation of depth of less than 18% of ocean beds has been obtained with a resolution of approximately a kilometer, using an echosounder (Mayer et al. 2018). A fortiori, a very small fraction has been inventoried, with a view to documenting the geological and biological diversity.

Scientific exploration of the fauna inhabiting the ocean bed began with some fortuitous discoveries. For example, we can cite the azoic hypothesis, proposed by Edward Forbes in the 19th century (Anderson and Rice 2006). He observed that the abundance of organisms, collected by fishing equipment up to a depth of around 500 meters, decreased with increasing depth. By statistical extrapolation, he hypothesized that there should be no

---

Chapter written by Sarah SAMADI and Sophie BARY.

organisms present beyond a depth of around 600 meters. The formulation of this hypothesis led to the collection of other data: contingently, such as the work of Milne-Edwards, who studied the animals collected along the transoceanic cables, which lived at depths of over 1,000 meters; or in an organized fashion, such as the scientific exploration of the world's oceans, notably with the iconic *Challenger* expedition, between 1872 and 1876. During this expedition, the onboard scientists took physical measurements (for example, depths and temperatures) and collected numerous geological and biological samples. These data showed that there are mountains and valleys on the ocean floor, and that whatever the depth, it is populated by a diverse range of organisms, though usually not present in large numbers. During the 20th century, the rollout of new techniques such as echo sounding allowed for more detailed mapping of undersea landscapes. The development of submersible machines enabled images of these landscapes to be captured. The exploration of the large abyssal plains confirmed that there, life is fairly sparse, but that there are "oases" of greater diversity within these deserts. The most striking discovery of such oases, to-date, has been the discovery of hydrothermal vents in the late 1970s: the hydrothermal vents, which lie along the oceanic ridges, release geothermally heated fluids, rich in chemical elements, which tend to be toxic to organisms; nevertheless, these areas are populated by dense fauna, with organisms which, at first glance, are very different to those encountered in other deep-ocean habitats.

In this chapter, looking at the *Tropical Deep-Sea Benthos* campaign, we shall illustrate how, over recent decades, our knowledge of the organisms in the upper strata of the ocean depths has continued to progress, and how the dynamics of that exploration has been influenced by the scientific, political and societal context.

2.2. The Tropical Deep-Sea Benthos program

2.2.1. *From MUSORSTOM to Tropical Deep-Sea Benthos*

In 1976, France's *Office de la recherche scientifique et technique outre-mer* (ORSTOM – Office for Overseas Scientific Research and Technology) – which was to become the *Institut de recherche pour le développement* (IRD – Institute of Research for Development) – decided to send its research vessel, the *Vauban*, to its center in Nouméa. On its way from Marseille to New Caledonia, the ship took a detour via the Philippines to conduct a

10-day oceanographic campaign, searching for a crustacean described as a living fossil. On 22 March 1976, the *Vauban* found *Neoglyphea inopinata* in exactly the spot where the *Albatross* had collected the first specimen, in 1908, off Lubang Island at 190 meters depth. Along with this spectacular rediscovery came a host of samples in all zoological groups. Jacques Forest and Alain Crosnier, the proponents of the campaign known as MUSORSTOM, to encourage collaboration between the Musée national d'Histoire naturel (MNHN) and ORSTOM, surrounded themselves with a network of experts to study them, describe them and publish about them; this marked the beginning of the MUSORSTOM campaigns, which, in 1999, became *Tropical Deep-Sea Benthos* (TDSB). Almost half a century later, the TDSB program has carried out over 7,500 dredging and trawling operations (see Table 2.1) in one of the last frontiers for biodiversity exploration: the bathyal zone of the largest tropical islands.

By analyzing the campaigns carried out as part of this program, between 1976 and 2016 (Bary 2018), it is possible to draw up a contemporary portrait of the issues at stake in naturalist exploration of marine biodiversity in the 21st century.

2.2.2. *TDSB campaigns across time and space*

The TDSB program is not a formal program, but follows in the footsteps of the oceanographic campaigns organized by a small group of researchers from the ORSTOM/IRD and the MNHN which, over time, expanded to include researchers from the CNRS, but also colleagues from universities in southeast Asia. The unifying factor between these campaigns is the scientific objective to explore the diversity of the benthic fauna in the bathyal zone of the largest tropical islands, the methods of sample collection and preservation, and the creation of networks of specialists spanning the globe to make use of the material collected. Figure 2.1 shows the position of these campaigns on a map centered on the Pacific.

With the launch of the research vessel *Alis* in 1986, the program entered into an unparalleled golden age in New Caledonia. Then, from 1992, its reach expanded into the archipelagos of the South Pacific, under the tireless leadership of Bertrand Richer de Forges. In the 2000s, it was the turn of the major archipelagos of Melanesia, with the Solomon Islands and Papua New Guinea. Partnerships were formed with other scientific teams, to use local vessels to explore the islands of Taiwan, the Philippines, Madagascar and the Mozambique Channel. In 2014, the program was further extended to the tropical Atlantic.

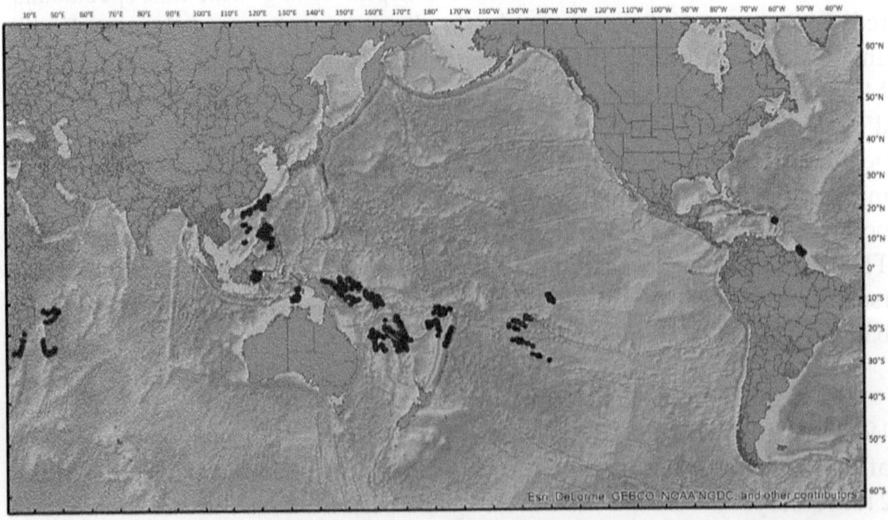

Figure 2.1. *Geographic distribution of campaigns run between 1976 and 2021*[1]. *For a color version of this figure, see www.iste.co.uk/nicolas/biodiversity.zip*

Table 2.1 lists all campaigns attached to the TDSB program, showing the year in which they took place, the acronym, the names of the mission leaders, the geographical area explored, the number of sampling operations, and, in the case of campaigns conducted on French research vessels, their DOI (Digital Object Identifier) in the database of the French oceanographic fleet[2]. Across the program as a whole, the most prominent mission leaders were Bertrand Richer de Forges (BRdF, 29), Sarah Samadi (SS, 9), Laure Corbari (LC, 7), Tin-Yam Chan (TYC, 7), Philippe Bouchet (PB, 6), Cécile Debitus (CD, 6), René Grandperrin (RG, 5) and Wei-Jen Chen (WJC, 5). With the exception of Cécile Debitus, who is a natural products chemist, and René Granperrin, who is a fisheries scientist, all other mission leaders have been zoologists or evolutionary biologists.

1 For further details, see the dedicated pages for each campaign, which can be found on: www.expeditions.mnhn.fr.

2 See: www.campagnes.flotteoceanographique.fr.

Date	Acronym	Mission leader(s)	Area	Op	doi
1976	MUSORSTOM1	J. Forest	Phi	59	10.17600/76380010
1979	VAUBAN78-79	C. Lévi	N-C	48	–
1980	CORINDON2	J. Launay	Ind	59	10.17600/80005712
1980	MUSORSTOM2	J. Forest	P	70	10.17600/80010211
1984	CHALCAL1	BRdF	N-C	98	10.17600/84001711
1985	BIOCAL	C. Lévi	N-C	68	10.17600/85002911
1985	MUSORSTOM3	J. Forest	Phi	59	10.17600/85005911
1985	MUSORSTOM4	BRdF	N-C	104	10.17600/85009111
1986	CHALCAL2	BRdF	N-C	38	10.17600/86006511
1986	MUSORSTOM5	BRdF	N-C	137	10.17600/86006611
1986	SMIB1	C. Debitus	N-C	15	10.17600/86009711
1986	SMIB2	C. Debitus	N-C	31	10.17600/86009611
1987	BIOGEOCAL	P. Cotillon	N-C	49	10.17600/87001811
1987	SMIB3	BRdF	N-C	33	–
1988	CORAIL2	BRdF	N-C	173	–
1989	CALSUB	M. Roux	N-C	22	10.17600/89009911
1989	GEMINI	M. Monzier	Van	13	10.17600/89005211
1989	MUSORSTOM6	BRdF	N-C	103	10.17600/89004811
1989	SMIB4	C. Debitus	N-C	36	10.17600/89004911
1989	SMIB5	C. Debitus	N-C	36	–
1989	VOLSMAR	M. Monzier	N-C	38	–
1990	AZTEQUE	RG	N-C	11	10.17600/90010211
1990	SMIB6	C. Debitus	N-C	40	10.17600/90005911
1991	BERYX2	RG	N-C	17	10.17600/91000811
1991	KARUBAR	MK. Moosa	Ind	91	–
1992	BERYX11	P. Lehodey	N-C	60	10.17600/92005011
1992	MUSORSTOM7	BRdF	W- F	145	10.17600/92005111
1993	BATHUS1	BRdF	N-C	75	10.17600/93000350
1993	BATHUS2	BRdF	N-C	58	10.17600/93000360
1993	BATHUS3	BRdF	N-C	78	10.17600/93000370
1993	LAGON	BRdF	N-C	NA	–
1993	SMIB8	C. Debitus	N-C	63	10.17600/93000640

1993	SMIB9	J.-L. Menou	N-C	NA	10.17600/93000650
1994	BATHUS4	BRdF	N-C	74	10.17600/94100030
1994	HALICAL1	RG	N-C	38	10.17600/94100020
1994	HALIPRO1	RG	N-C	32	10.17600/94100010
1994	MUSORSTOM8	BRdF	Fidji	186	10.17600/94100040
1995	SMIB10	?	N-C	15	–
1996	HALIPRO2	RG	N-C	106	10.17600/94100010
1996	SMCB	J. Poupin	FP	590	–
1997	MUSORSTOM9	BRdF	FP	165	10.17600/97100020
1998	MUSORSTOM10	BRdF	Fidji	83	10.17600/98100080
1999	BORDAU1	BRdF	Fidji	117	10.17600/99100020
1999	LITHIST	BRdF	N-C	18	–
1999	PALEO-SURPRISE	BRdF	N-C	28	10.17600/99100040
2000	BORDAU2	BRdF	Tong	111	10.17600/100060
2000	TAIWAN2000	TYC	Tw	55	–
2001	NORFOLK1	BRdF	N-C	89	10.17600/1100050
2001	SALOMON1	BRdF	Sol	118	10.17600/1100090
2001	TAIWAN2001	TYC	Tw	80	–
2002	BENTHAUS	BRdF	FP	318	10.17600/2100100
2002	TAIWAN2002	TYC	Tw	45	–
2003	NORFOLK2	BRdF	N-C	141	10.17600/3100030
2003	TAIWAN2003	TYC	Tw	24	–
2004	BOA0	BRdF	Van	27	10.17600/4100140
2004	SALOMON2	PB	Sol	141	10.17600/4100090
2004	TAIWAN2004	TYC	Tw	34	–
2005	BOA1	SS BRdF	Van	71	10.17600/5100060
2005	EBISCO	BRdF	N-C	172	10.17600/5100080
2005	PANGLAO2005	PB L Labe PKL Ng	Phi	79	–
2006	SANTO2006	SS BRdF	Van	134	10.17600/6100100 10.17600/6100110
2007	AURORA2007	PB MM Santos	Phi	224	–
2007	SALOMONBOA3	BRdF	Sol	96	10.17600/7100070
2008	CONCALIS	PB	N-C	103	10.17600/8100010

2008	LUMIWAN2008	PB	Phi	68	–
2008	TERRASSES	SS	N-C	99	10.17600/8100100
2009	MAINBAZA	PB A Ramos	SwIO	46	–
2009	MIRIKY	PB	SwIO	119	–
2009	TARASOC	PB	FP	212	10.17600/9100040
2010	ATIMOVATAE	PB	SwIO	121	–
2010	BIOPAPUA	SS LC	PNG	156	10.17600/10100040
2011	EXBODI	SS LC	N-C	161	10.17600/11100080
2012	PAPUANIUGINI	SS	PNG	265	10.17600/18000841
2013	TAIWAN2013	TYC WJC	Tw	16	–
2014	DONGSHA2014	WJC	Tw	12	–
2014	GUYANE2014	PB LC	FG	68	–
2014	KAVIENG2014	PB	PNG	95	10.17600/14004400
2014	MADEEP	SS LC K Olu	PNG	105	10.17600/14004000
2014	NANHAI2014	WJC	Tw	21	–
2015	KARUBENTHOS2	PB LC	FA	139	10.17600/15005400
2015	ZHONGSHA2015	W.-J. Chen	Tw	36	–
2016	KANACONO	SS N Puillandre	N-C	134	10.17600/16003900
2017	BIOMAGLO	SS LC K Olu	SwIO	89	10.17600/17004000
2017	KANADEEP1	SS S Hourdez	N-C	139	10.17600/17003800
2018	MD208	PB J-F Ternon LC	SwIO	56	10.17600/17002700
2018	KAVALAN	WJC	Tw	14	–
2019	KANADEEP2	SS K Olu	N-C	266	10.17600/18000883
2019	SAKIZAYA	WJC	Tw	25	–
2021	SPANBIOS	SS	N-C	180	10.17600/18000701

Table 2.1. *List of TDSB campaigns from 1976 to 2021. The names of the most prominent mission leaders have been replaced by their initials*

COMMENT ON TABLE 2.1.– *Geographic areas explored: Phi = Philippines; Ind = Indonesia; N-C = New Caledonia; Sol = Solomon Islands; PNG = Papua New Guinea; SwIO = South-west Indian Ocean; W-F = Wallis and Futuna; FP = French Polynesia; FG = French Guyana; FA = French Antilles; Van= Vanuatu; Tw = Taiwan. Op = Number of fauna sampling operations carried out.*

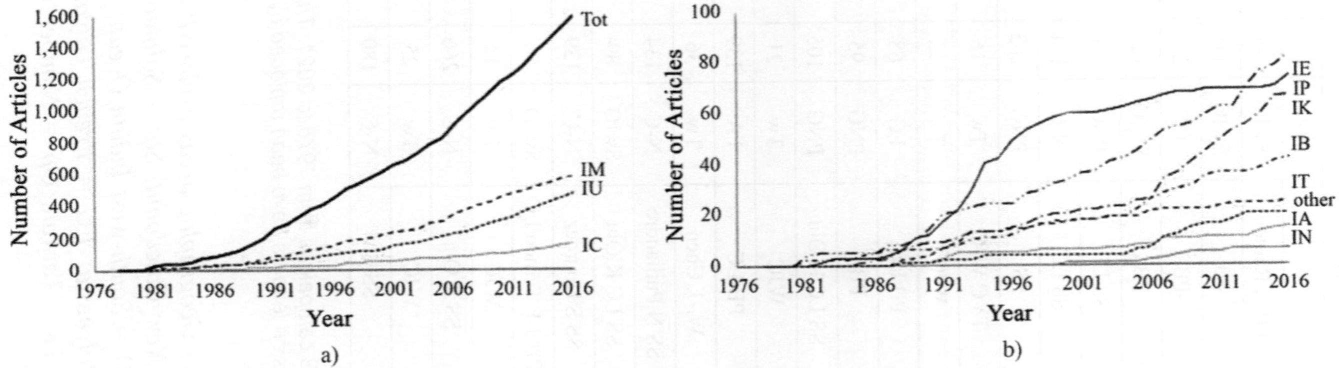

Figure 2.2. *Cumulative curves showing the number of publications per year for each of the zoological groups corresponding to the MNHN collection codes*

COMMENT ON FIGURE 2.2.– *a) Total curve on which IM = mollusks; IU = crustaceans; IC = ichthyology. b) IA = annelids-polychaetes, IB = brachiopods-bryozoa, IE= echinoderms; IK = cnidaria; IN = nematodes; IP = sponges; IT = tunicata-ascidiacea. Other = single-celled organisms or fossils.*

2.2.3. *Program metrics*

A systematic census of all scientific articles, having studied samples or data collected during campaigns run since 1976, is carried out on a regular basis. The results are available on the dedicated pages of the MNHN's expeditions database[3]. Between 1976 and 2016, Sophie Bary counted 1,579 articles, and showed that the majority of these articles deal with taxonomy or systematics. Of these articles, 63% include descriptions of new taxa. The different taxonomic groups are not equally represented: 36% of articles relate to mollusks, 30% to crustaceans, and 10% to ichthyological collections. For each of the other taxonomic groups, there are fewer than 100 articles (see Figure 2.2).

Another interesting metric is the number of taxa described. In total, in the space of 40 years, this campaign program has led to the description of 3,628 species; of these, 44% are mollusks and 34% were crustaceans. Publications linked to ichthyological collections, which represent 10% of articles, provide only 4% of the total number of taxa newly described.

2.3. Continuity and ruptures in the TDSB program

2.3.1. *The place of "living fossils" in discourse on the deep oceans*

The azoic hypothesis developed by Edward Forbes and the discovery of specimens assigned to taxa in the fossil record marked the scientific representation of the ocean depths for over a century. The reports from the first three MUSORSTOM campaigns are testament to this fact, with their main scientific justification being the search for the decapod crustacean *Neoglyphea inopinata*. This species, which was described in 1975 by Jacques Forest and Michèle Saint Laurent, taxonomists from the MNHN,

3 See: www.expeditions.mnhn.fr.

was attributed by those authors to the Glypheidea line, which was assumed to have been extinct for over 50 million years. The history of the search for *Neoglyphea* plays a crucial part in the way in which certain researchers involved in the program tell its story (Forest 1981; Bouchet et al. 2008; Richer de Forges et al. 2013), giving it the status of a founding myth.

Since the beginnings of the exploration of the deep seabed in the 19th century, zoologists hypothesized that "living fossils" could be found in the ocean depths. During explorations carried out by the naturalists Georg O. Sars and his son along the Norwegian coasts, a species of peduncular crinoid *Rhizocrinus lofotensis*, Sars 1868, was collected at 550 meters in 1864. Sars and Sars (1872) compared that organism to a fossil shape that was thought to be extinct, describing it as a "living fossil". However, though the term "living fossil" is used in reports on (and applications for) TDSB campaigns, the term is rarely found in the scientific literature in which subsequent study is made of the samples collected during those campaigns. Between 1976 and 2016, this term appeared in fewer than 2% of the 1,541 articles examined. These shapes, which resuscitate lineages previously known before only from fossil records, enable advances to be made in phylogenetic history, as they can be included in molecular phylogeny studies.

As Gage and Tyler (1991, p. 12) highlight, certain scientific opinions on the deep oceans endured for longer than they ought to have done, because of the scarcity of data. Thus, the repeated occurrence of the term "living fossil" in publications around seabed exploration campaigns is attributable to lack of knowledge. Paradoxically, the fossil record of the ancient oceans which is now on land is better known, because it is easier to explore than the fauna currently living in the depths of the oceans: thus, it is tempting to consider present forms as ancient, simply because the forms of the past were discovered before those currently living were.

The last major discovery of these lines, which today are described as relicts rather than as "living fossils" was the discovery of a single specimen of a new family of Ophiuroidea which, according to molecular dating, appears to have diverged from other forms found today, around the time of the Jurassic. Evidence of the ancientness of this line can be found in its similarity to microfossils from the Jurassic (O'Hara et al. 2021).

2.3.2. *A program evolving with societal issues*

Analyzing the evolution of the communities of researchers at sea and in scientific publications illustrates the transitions over the period 1976–2016. Over 40 years, we have seen a generational change, in parallel to the expansion of the zones being explored and the institutions and nations involved (see Table 2.1 and Figure 2.1, and the analyses presented in Bary 2018). Before the turn of the millennium, mission leaders and participants in the campaigns tended, for the most part, to be of French origin, and exploration was focused around New Caledonia. The dawn of the 21st century brought a geographic expansion of the zones being explored and the institutions involved.

From the standpoint of the objectives, it can be seen that, prior to the 2000s, a significant portion of the campaigns, carried out primarily in New Caledonia, had other declared objectives than the exploration of faunistic diversity. For example, the program *Substances marines d'intérêt biologique* (SMIB – Marine Substances of Biological Interest), led mainly by Cécile Debitus, comprised some ten campaigns which, in quantitative terms, produced more taxonomic results than results in chemistry (see Figure 2.3a). Numerous campaigns during this period also had objectives related to the evaluation of fisheries resources, though they were indistinguishable from other campaigns from the standpoint of their scientific production. These declarations show how important it was, during this period, to justify the existence of these campaigns by pointing to what they contributed in the race for new resources in these overseas territories.

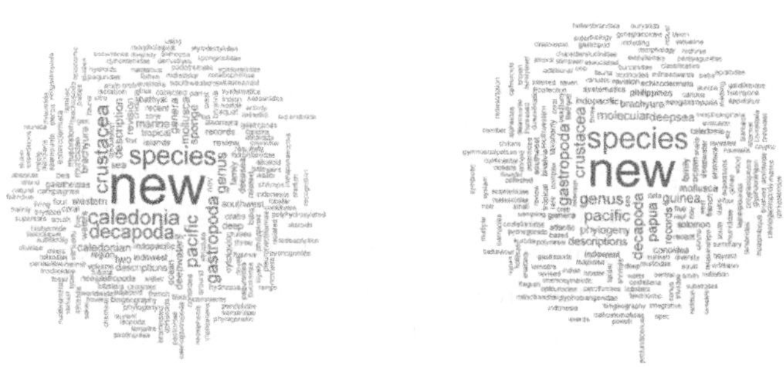

a) b)

Figure 2.3. *Word cloud created from the titles of the 348 articles published a) on the basis of samples taken during the SMIB campaigns (1986–1995); b) on the basis of the titles of the 154 articles published between 2010 and 2016, using collections from the campaigns in Papua New Guinea between 2010 and 2014. For a color version of this figure, see www.iste.co.uk/nicolas/biodiversity.zip*

This situation changed at the dawn of the 21st century, with campaigns clearly focusing on obtaining data on biodiversity and recognition of the need for knowledge in relation to the questions regarding anthropic impacts (pollution, climate change, shocks, mineral resources, etc.) in these inaccessible environments. Figure 2.3b, looking at the period 2010–2016, illustrates the link between the mission objectives declared in campaign applications and the publications for the campaigns carried out in Papua New Guinea. Among the publications resulting from these campaigns in Papua New Guinea, the article by Samadi et al. (2015), for example, illustrates the way in which the results on biodiversity are mobilized to answer questions relating to the anthropic impact on biodiversity in the deep oceans. This cloud also highlights the more important role played by results in molecular systematics, shown by the appearance of words such as "phylogeny" and "molecular" in article titles – words which were completely absent from the word cloud produced from the SMIB campaigns. The two word clouds shown in Figure 2.3 demonstrate that these naturalistic collections, which enriched the collections of the National Museum of Natural History in Paris, fed into the research projects whose objectives were not necessarily those which were used to justify the execution of the campaigns.

2.4. Campaigns at sea with taxonomic results: a network dynamic around naturalist collections

The success of the TDSB program rests on clearly defined sampling methods, the constitution of naturalist collections available through the naturalist collection infrastructures (primarily, the MNHN in Paris) and the running of a broad network of specialists, both amateurs and professionals, capable of studying these collections whilst taking on board the evolution in taxonomic investigation techniques. Today, campaigns, carried out over 40 years ago, continue to provide us with specimens and new species. We could point to the example of the BIOCAL campaign in 1985, carried out in New Caledonia, the samples from which continue to supply data for major publications on deep-sea biodiversity in that region. One example of such a publication is the monograph on New Caledonia's deep corals, published by Kitahara and Cairns (2021), and which, using the material collected during 49 campaigns, identified 267 species, 47 of which were new to science.

For over 45 years, the study of the collections resulting from these campaigns has occupied over 200 scientists all over the world, who have lauded the exceptional quality and diversity of the material sampled. Hundreds of scientific articles, 33 volumes in the eponymous series Tropical Deep-Sea Benthos, support provided to guest researchers, theses, the involvement of high-level amateurs: systematics of marine organisms has reaped the benefit of the remarkable collective drive of a program which has drawn on the knowledge of an entire chain of actors, boat crews, onboard scientists, collection technicians, researchers and writing assistants. In addition to the discovery of thousands of new species, Tropical Deep-Sea Benthos was also pioneering in the study of the fauna associated with wood debris (for example, see Lee et al. 2022). The project explored the endemism and vulnerability of seamounts (for example, see Richer de Forges et al. 2000; Pante et al. 2015; Delavenne et al. 2019). It has been the main data source concerning the biodiversity of the benthos, allowing for policies to be put in place, aimed at conserving the tropical high-sea domain (for example, see Castelin et al. 2019).

Analysis of the scientific production by participants in the campaigns shows that, the longer these people spend at sea, the more they tend to contribute to the program's scientific production (Bary 2018). This analysis highlights the importance of the mission leaders in promoting these activities, but especially the particular role of Alain Crosnier, who had a major influence in driving forward the networks of specialists. This role is highlighted in the article by Macpherson et al. (1998), written in tribute to his work. Alain Crosnier, who died at the age of 91 in 2021, had valorization of the data at heart (Richer de Forges et al. 2021), to such an extent that he made a donation to the MNHN to support the inviting of specialists, amateurs or retired academics who could not be supported by institutional funds to continue their work. The efforts continue even today, with a new generation of researchers, using new techniques in taxonomy, but also in terms of the means employed to collect samples.

2.5. References

Anderson, T.R. and Rice, T. (2006). Deserts on the sea floor: Edward Forbes and his azoic hypothesis for a lifeless deep ocean. *Endeavour*, 30(4), 131–137.

Bary, S. (2018). Les représentations de la biodiversité dans les fonds marins : une approche épistémologique et scientifique. PhD Thesis, Université Paris Descartes, Paris.

Bouchet, P., Héros, V., Lozouet, P., Maestrati, P. (2008). A quarter-century of deep-sea malacological exploration in the South and West Pacific: Where do we stand? How far to go? In *Tropical Deep-Sea Benthos 25*, Héros, V., Cowie, R.H., Bouchet, P. (eds). Muséum national d'Histoire naturelle, Paris, 9–40.

Castelin, M., Delavenne, J., Brisset, J., Chambart, C., Corbari, L., Keszler, L., Lozouet, P., Olu, K., Poncet, L., Puillandre, N. et al. (2019). Exploration et étude de la biodiversité benthique profonde de Mayotte et des îles éparses. Technical Report, MNHN.

Delavenne, J., Keszler, L., Castelin, M., Lozouet, P., Maestrati, P., Samadi, S. (2019). Deep-sea benthic communities in the largest oceanic desert are structured by the presence of polymetallic crust. *Scientific Reports*, 9(1), 6977.

Forest, J. (1981). Compte rendu et remarques générales. In *Résultats des campagnes MUSORSTOM I. Philippines (18–28 mars 1976)*, Forest, J. (ed.). ORSTOM, Paris, 9–50.

Gage, J.D. and Tyler, P.A. (1991). *Deep-Sea Biology: A Natural History of Organisms at the Deep-Sea Floor*. Cambridge University Press, Cambridge.

Gouvernement de la République Française (2023). Explorer les grands fonds marins pour mieux les protéger [Online]. Available at: https://www.gouvernement.fr/actualite/explorer-les-grands-fonds-marins-pour-mieux-les-proteger [Accessed 19 July 2023].

Kitahara, M.V. and Cairns, S.D. (2021). *Tropical Deep-Sea Benthos 32: Azooxanthellate Scleractinia (Cnidaria, Anthozoa) from New Caledonia*. Muséum national d'Histoire naturelle, Paris.

Lee, H., Puillandre, N., Kano, Y., Chen, W.J., Samadi, S. (2022). Biodiversity and phylogeny of Cocculinidae (Gastropoda: Cocculinida) in the Indo-West Pacific. *Zool. J. Linn. Soc.*, 196(1), 366–392.

Macpherson, E., Lemaitre, R., Richer de Forges, B., Manning, R.B. (1998). Crosnier's cronies and their view from the attic window. *Zoosystema*, 20, 139–142.

Mayer, L., Jakobsson, M., Allen, G., Dorschel, B., Falconer, R., Ferrini, V., Lamarche, G., Snaith, H., Weatherall, P. (2018). The Nippon Foundation–GEBCO Seabed 2030 project: The quest to see the world's oceans completely mapped by 2030. *Geosciences*, 8(2), 63.

O'Hara, T.D., Thuy, B., Hugall, A.F. (2021). Relict from the Jurassic: New family of brittle-stars from a New Caledonian seamount. *Proc. R. Soc. B.*, 288, 20210684.

Pante, E., France, S.C., Gey, D., Cruaud, C., Samadi, S. (2015). An inter-ocean comparison of coral endemism on seamounts: The case of *Chrysogorgia*. *Journal of Biogeography*, 42(10), 1907–1918.

Richer de Forges, B., Koslow, J.A., Poore, G.C.B. (2000). Diversity and endemism of the benthic seamount fauna in the southwest Pacific. *Nature*, 405(6789), 944–947.

Richer de Forges, B., Chan, T.-Y., Corbari, L., Lemaitre, R., Macpherson, E., Ahyong, S.T., Ng, P.K.L. (2013). The MUSORSTOM-TDSB deep-sea benthos exploration programme (1976–2012): An overview of crustacean discoveries and new perspectives on deep-sea zoology and biogeography. In *Tropical Deep-Sea Benthos 27*, Ahyong, S.T., Chan, T.-Y., Corbari, L., Ng, P.K.L. (eds). Muséum national d'Histoire naturelle, Paris, 13–66.

Richer de Forges, B., Ahyong, S.T., Castro, P., Chan, T.Y., Clark, P.F., Lemaitre, R., Tavares, M. (2021). Alain Crosnier's role in modern carcinology: Exploration, international collaboration, and taxonomy. *J. Crust. Biol.*, 41(3), ruab031.

Samadi, S., Puillandre, N., Pante, E., Boisselier, M.C., Corbari, L., Chen, W.J., Hourdez, S. (2015). Patchiness of deep-sea communities in Papua New Guinea and potential susceptibility to anthropogenic disturbances illustrated by seep organisms. *Mar. Ecol.*, 36, 109–132.

Sars, G.O. and Sars, M. (1872). *On Some Remarkable Forms of Animal Life from the Great Deeps off the Norwegian Coast*, volume 1. Brøgger & Christie, Christiana.

Secrétariat d'état chargé de la mer (2023). La connaissance de l'océan sera enrichie par l'objectif "Grands fonds marins" de France 2030 [Online]. Available at: https://mer.gouv.fr/la-connaissance-de-locean-sera-enrichie-par-lobjectif-grands-fonds-marins-de-france-2030 [Accessed 6 July 2023].

3

CEAMARC: An Integrated Campaign to Evaluate Biodiversity at All Scales in Adélie Land

3.1. Introduction

Antarctica is, in our collective imaginaries, one of the last places on Earth where nature is still the undisputed king. This particular construct of "nature" is at the crossroads between tales of sealers and whalers, explorers, scientists, and diplomats and geopolitics (Fogg 1992).

In 1961, the Antarctic Treaty, signed in 1959 by 12 countries, including the United States of America and the Soviet Union, entered into force, freezing the territorial claims previously asserted by the States. This treaty stipulates that "Antarctica shall be used for peaceful purposes only" (article 1, paragraph 1) and that "freedom of scientific investigation in Antarctica and cooperation toward that end, as applied during the International Geophysical Year, shall continue" (article 2). Over the course of the decades, the Antarctic Treaty has been supplemented by additional treaties: the Convention for the Conservation of Antarctic Seals (CCAS) in 1972; the Convention for the Conservation of Antarctic Marine Living Resources (CCAMLR) in 1980; and finally, the Madrid Protocol, in 1991, which consecrated the Antarctic as a nature reserve, devoted to peace and

Chapter written by Marc ELÉAUME, Nadia AMÉZIANE, Frédéric BUSSON, Romain CAUSSE, Agnès DETTAÏ, Guillaume LECOINTRE and Catherine OZOUF-COSTAZ.

science. This system of treaties makes the Antarctic and the Southern Ocean a unique place in the world: a haven for peace, yet an immense laboratory for studying global change.

France is one of the seven sovereign states which agreed to suspend their territorial claims in Antarctica. The part claimed by France, Adélie Land (see Figure 3.1), is a small zone of around 500,000 square kilometers, framed by two vast sectors claimed by Australia. Adélie Land is a tiny portion of eastern Antarctica: a frozen world, blasted by katabatic winds, where Jules Dumont d'Urville disembarked one afternoon in January 1840, and claimed possession in the name of France, right under the nose of Charles Wilkes, whom he encountered only a week later. It would be more than a century before the *Expéditions polaires françaises* (EPF, French Polar Expeditions) installed the first station at Port-Martin, and subsequently on Petrel Island, where it became a permanent research station in 1959.

The research carried out in Adélie Land has greatly contributed to our knowledge of the dynamics of the ice cap, the sea ice, and the populations of higher predators (birds and marine mammals). Since the 2000s, sensors have been fitted to the wildlife, which have been used to measure the essential variables of the ocean, both summer and winter. They are also considered to be sentinels of the state of the ecosystems, and thus are an essential element in the CCAMLR ecosystem monitoring program. This monitoring provides information about the effects of natural variations in the environment in the absence of fishing, and, where fishing is practised, helps distinguish the effects of natural variations from the effects of anthropic pressure on ecosystems.

The results of this research have allowed the CCAMLR to quantify the urgency of the need to implement efficient tools for ecosystem management. Aside from the conservation measures which govern human activities in the Southern Ocean, the CCAMLR decided to create a representative system of marine protected areas (MPAs), to contribute to the resilience of flora and fauna in the Southern Ocean in the context of rapid climate change and increasing fishing activities.

3.2. The CEAMARC expedition (Collaborative East-Antarctic Marine Census)

3.2.1. *An "integrated" campaign*

The aim of the *Census of Antarctic Marine Life* (CAML[1]) was to understand marine biodiversity in Antarctica in all its aspects, in order to gain better knowledge on Southern Ocean ecosystem functioning. The CAML achieved the feat of setting up an international consortium to study Antarctica's marine biodiversity in an integral manner. The International Polar Year (IPY 2007–2008; see: www.ipy.org) was indubitably the acme of the CAML program, with scientists from 30 countries and 50 institutions, aboard 14 research vessels, working cooperatively all around Antarctica, to produce a point of reference for biodiversity in the Southern Ocean (Ausubel et al. 2010).

In the case of eastern Antarctica, this challenge was taken up by Australia, France, Japan and Belgium, in the context of the CEAMARC expedition (*Collaborative East-Antarctic Marine Census*), during the 2007–2008 austral summer, on board three icebreaker research vessels, the SV *Astrolabe*, RV *Umitaka Maru* and RSV *Aurora Australis* (Hosie et al. 2011). The expedition focused on the continental shelf, the continental slope and the open sea off the coast of Adélie Land and George V Land (see Figure 3.1), between longitudes 138°E (a little to the west of the Astrolabe Glacier and France's Dumont d'Urville station) and 148°E (the Mertz Glacier) on seabeds ranging from 100 meters to over 2,000 meters depth (Beaman and O'Brien 2009). The sampling efforts were distributed among the vessels, on the basis of the habitat being studied. Each ship explored a particular aspect of the biodiversity: the *Aurora Australis* collected benthic organisms and demersal fish from the continental shelf and the slope down to 2,000 meters. The *Umitaka Maru* took samples of plankton, pelagic and mesopelagic fish on the shelf and in the open sea, and the *Astrolabe* sampled plankton in the coastal region.

The strategy of random systematic sampling was chosen, due to the limited available knowledge in the area being studied (in comparison to

1 Available at: www.caml.aq.

other regions of the globe), though numerous previous campaigns and research programs had provided clues. At the meso-scale, the data collected by the *Australasian Antarctic Expedition* (1911–1914), the *British, Australian and New Zealand Antarctic Research Expedition* (BANZARE; 1929–1931) and the WEGA project, on board the *Tangaroa* (2000) identified contrasting geographic zones, allowing for the description of the diversity of benthic habitats in the region (Beaman and Harris 2003, 2005). On a smaller scale, in 1961, Patrick Arnaud, during his overwintering at Dumont d'Urville, began to describe the benthic and demersal communities down to 150 meters depth around the Dumont d'Urville station (Hureau 1962; Hureau and Arnaud 1964; Arnaud 1974). Launched in 1996, the program ICOTA (Ichtyologie côtière en Terre Adélie (Coastal Ichthyology in Adélie Land)) run by the Institut Paul Émile Victor, also produced a more detailed inventory of coastal notothenioid fish around Adélie Land to a depth of 200 meters, and the characteristics of their biotopes (Gutt et al. 2007; Koubbi et al. 2010).

Aurora Australis collected samples between 22 December 2007 and 19 January 2008, during the Austral summer, which is the only time of year when the melting of sea ice allows for exploration of the continental shelf of the Dumont d'Urville Sea. In total, 81 stations were explored, and benthic samples (trawling or dredging) (see Figures 3.2a and 3.2b), sediment samples (box corers, Van Veen grab sampler, Smith-McIntyre grab sampler) (see Figure 3.2c), and water samples (Niskin bottles) (see Figure 3.2d) were collected from each of them. Photo and video transects were also carried out, along with measurements of conductivity, temperature and depth (CTD), current speed and direction (ADCP), bathymetry (hull-mounted monobeam acoustic sounder).

The CEAMARC expedition is an example of an integrated campaign which has been able to explore benthic populations in a poorly known region, and to measure the environmental factors which account for the distribution of the communities. The biodiversity in the area was studied from multiple angles, from species richness to models of ecological niches, providing a detailed understanding of the ecosystems. This approach established a point of reference which, today, constitutes the baseline data underlying the East-Antarctic MPA proposal.

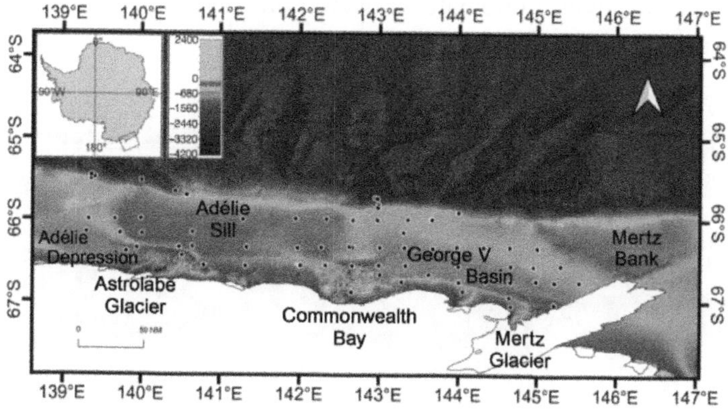

Figure 3.1. *Map of the sea around Dumont d'Urville, including the main toponyms. The color gradient indicates depth; the black dots mark the sampling stations. For a color version of this figure, see www.iste.co.uk/nicolas/biodiversity.zip*

3.2.2. *Sampling methods*

3.2.2.1. *Beam trawl*

The beam trawl (see Figure 3.2a) comprises a beam – which can break if it snags on the sea bed – affixed to two metal "shoes". This ensemble forms a rigid structure, to which the trawl (net) is attached. Underneath the net, chains are attached. These rake the seabed and lift up the benthic fauna, which is then collected in the net.

During the CEAMARC expedition, a beam trawl measuring 3 meters in width × 1.39 meters in height, with a 10mm mesh stretched over the mouth of the trawl was used. It was deployed on 87 occasions, over periods ranging from 5 to 26 minutes (depending on the depth), to collect benthic and demersal organisms in the area at depths ranging from 150 to 2,065 meters (Causse et al. 2011). When used on highly irregular seabeds, the trawl was replaced by a benthic sled (see Figure 3.2b). Most of the taxa were sorted and collected in their entirety, but the largest and most abundant (sponges, ascidians, holothurians, etc.) were subsampled. It is therefore possible to calculate abundance indices for certain taxa. Each sorted batch was associated with a unique number, and the number of specimens in the batch was recorded. Each batch, labeled with the details of the station where it was collected, was entered into a database, which was the forerunner of what, today, is the "référentiel des campagnes" (Repository of Surveys) run by the

MNHN[2]. In total, over 7,000 batches were collected, and sent to experts all over the world for identification.

Figure 3.2. *Some of the gear deployed during the CEAMARC expedition: a) a beam trawl equipped with a video camera or photo camera; b) a benthic sled; c) a box corer; d) a rosette station equipped with 24 Niskin bottles, a CTD and an ADCP. Copyright: AAD-MNHN. For a color version of this figure, see www.iste.co.uk/nicolas/biodiversity.zip*

The beam trawl was equipped with a high-definition video camera on 33 occasions, and a high-definition photo camera on 39 occasions. The images thus captured can be used to describe the benthic landscapes and the various communities which inhabit them. Some such communities, essentially comprising sessile filter feeders and habitat builders (e.g. sponges, ascidians, cnidarians, etc.), were identified as being indicators of vulnerable marine ecosystems (VMEs). These VMEs are subject to special conservation measures by the CCAMLR, and are now protected from any activity which could negatively impact these ecosystems. The transects of images captured also shed light on areas scoured by icebergs, and help to assess the

2 See: www.expeditions.mnhn.fr/.

ecological successions which come in the wake of these significant reshapings of the benthic landscape.

3.2.2.2. *Box corer, Van Veen and McIntyre grabs*

The box corer (see Figure 3.2c) is a sediment sampling system. It consists of a pair of jaws which are set in the open position at the moment of deployment. The corer descends under the influence of gravity, and the jaws snap shut when it hits the bottom, capturing a given volume of sediment. The corer is then retrieved, protecting the sediment from leaching as it is brought up. The sediment remains in place, so its surface and the associated fauna can be analyzed.

The box corer was used the most frequently to collect samples of sediments and of the associated endogenous and exogenous macrofauna. It collects samples of the top 10–20cm of sediment. The samples can be used to assess particle-size distribution, total organic-carbon content, total organic-nitrogen content, carbon and nitrogen isotopes, biogenic silica and carbonate content. In total, 102 box corers were deployed, and 306 samples were collected across 51 stations.

The sediments containing higher proportions of gravel and sand were more difficult to sample with a box corer. Van-Veen grabs (20 samples over 3 stations) and Smith-McIntyre grabs (4 samples over 2 stations) were used when the box corer came back up without sediments (Beaman and O'Brien 2009; Hosie et al. 2011).

3.2.2.3. *Rosettes, Niskin bottles and CTD and ADCP sensors*

The rosette (see Figure 3.2d) is a cylindrical metal structure onto which tools and sensors can be fixed. Typically, a rosette carries between 12 and 36 Niskin bottles and a CTD sensor. Niskin bottles are water-sampling tools. The bottles are attached in the open position on both sides, and closed when the desired depth is reached, by a signal carried by an electrical (messenger) cable.

A CTD device is a tool measuring conductivity, temperature and water pressure continuously throughout the water column. The CTD can be attached to a rosette, which allows a large number of tools to be deployed at the same time.

An ADCP (Acoustic Doppler Current Profiler) sends out an acoustic signal into the water column. The signal is backscattered by moving particles, and captured by a transducer, capable of calculating the direction and speed of the current.

The rosette and the components it carries are essential elements in physical oceanography. The CTD assesses the stratification of water masses, the ADCP analyzes current speed and direction, and water analyses can provide data on, for instance, the partial pressure of dissolved gases (O_2, CO_2, etc.), stable isotopes of carbon and nitrogen, trace elements, etc. Using these measurements, we can derive certain essential variables which restrict the distribution of marine flora and fauna. These variables can be used to describe the ecological niches inhabited by taxa of interest, and to calculate the probability of their being found in locations that have not been sampled.

During the CEAMARC campaign, the rosette was deployed no fewer than 130 times. In total, 24 Niskin bottles can be carried at once, so samples can be taken at 24 different depths during the same deployment. Measurements can be taken of salinity, dissolved oxygen, phosphate, nitrate, nitrite and silicate, ^{18}O, CFC, dissolved organic carbon, alkalinity, ^{14}C, germanium/silica/boron, isotopes of germanium, silicon and boron, trace metals, neodymium, chlorophyll a, cell counts and pigments (Rosenberg and Rintoul 2010).

3.2.2.4. *Monobeam sounder*

The *Aurora Australis* was fitted with a monobeam sounder which can be used to measure the height of water beneath the vessel's keel. The sounder sends an acoustic signal to the seabed, which is then reflected and captured by a transducer on the hull. The depth can be calculated by examining the signal's time-of-flight.

During the CEAMARC expedition, the echosounder was used to supplement the preexisting dataset and to much more precisely map the geomorphological structures in the area (Beaman et al. 2011).

3.3. Some results

The results stemming from the CEAMARC expedition are already numerous, with over 140 articles published over the past 15 years. Some of this work – particularly that based on robust systematics – has only been able to be done over the long term. It provides a clearer view of biodiversity in the Dumont d'Urville Sea at all scales of integration, from the single gene to the entire ecosystem.

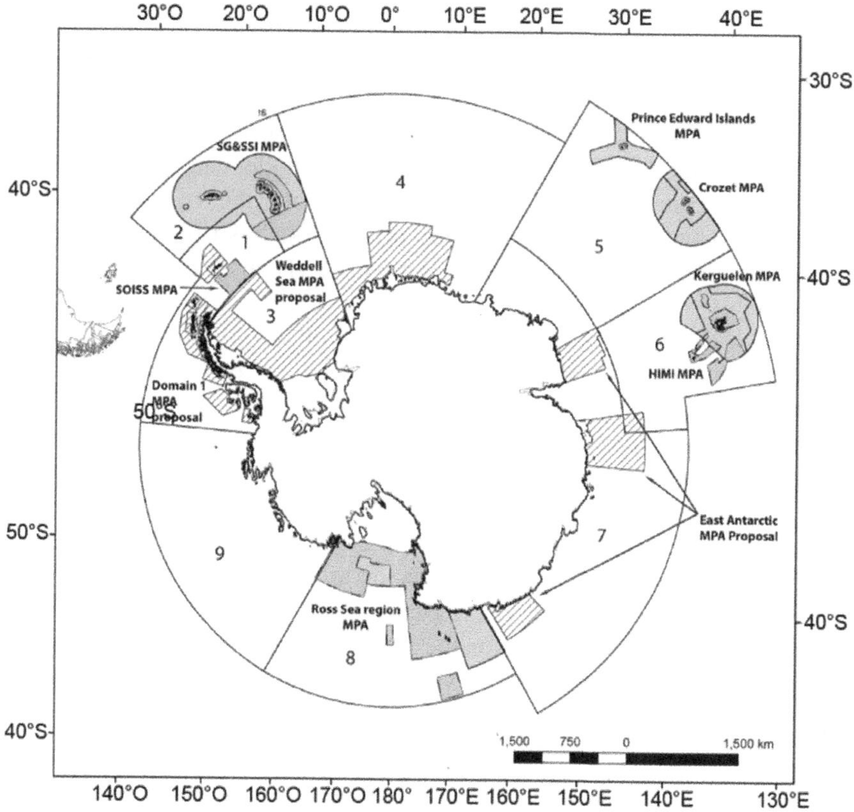

Figure 3.3. *Proposed areas for the CCAMLR and marine protected areas in the Southern Ocean. Modified from Brooks et al. (2020). For a color version of this figure, see www.iste.co.uk/nicolas/biodiversity.zip*

COMMENT ON FIGURE 3.3.– *The existing MPAs are colored gray (solid line), prohibited areas are colored blue (dotted lines) and the proposed MPAs are represented by hashed shading. SOISS is the South Orkney Islands Southern Shelf MPA; SG and SSI represent the South Georgia and South Sandwich Islands MPA; HIMI represents the Heard Island and McDonald Islands marine reserve.*

By analyzing the echosounder measurements, alongside those obtained by previous campaigns, it has been possible to describe the main morphological features of the seabed on the continental shelf of the Dumont d'Urville Sea (Beaman et al. 2011). These data are absolutely essential in understanding the structure of the benthic ecosystems: as necessary, in fact, as the classification of sediments across the whole region (Post et al. 2011). In addition to these fundamental data, we have a further dataset describing the biogenic silica and carbonate content of the seabed, and the characterization of the water masses. These data were fed into a regional oceanographic model (Galton-Fenzi 2009). This model can be used, for any point in the Dumont d'Urville Sea, to calculate the mean values and variances of the explicative variables driving the distribution of benthic communities, such as current speed and direction (Hemery et al. 2011).

The species inventory included over 7,000 cataloged batches, more than 17,000 specimens distributed across 13 phyla, over 300 genera and 340 species identified. Many specimens identified as species have been barcoded, enabling the constitution of a very exhaustive reference dataset for the region (Dettaï et al. 2011). Also, the genetic data produced have helped us to address key questions linked to the history of Antarctica: is the Southern Ocean a species flock (Lecointre et al. 2013; Chenuil et al. 2018)? Are species circumpolar in distribution, or are they complexes of distinct cryptic species (Allcock et al. 2011; Hemery et al. 2012)? Do glacial cycles in the Southern Ocean account for the profusion of cryptic species (Arango et al. 2011; Baird et al. 2011; Soler-Membrives et al. 2017)?

3.4. Conclusion

The CEAMARC zone is one of the three scientific reference areas identified for the East Antarctic MPA (see Figure 3.3) and will, ultimately, become part of the representative system of MPAs for the Southern Ocean. This zone covers a wide range of ecosystems, from the oceanic zone, through the continental shelf, to the coastal zone. The different ecoregions found to exist, thanks to the biotic and abiotic data collected through the CEAMARC program (Koubbi et al. 2010) are surprisingly close to previous attempts at regionalization (Beaman and Harris 2005; Grant et al. 2006) performed with independent datasets. They reinforce the zoning carried out all around Antarctica, with the goal of establishing a representative system of MPAs.

In addition, this work contributed to the construction of the management plan for the East-Antarctic MPA. As a result of the discovery of hyperdiverse ecoregions in the George V Basin, the area has been precluded from any potentially harmful human activity. Analysis of image transects and water samples led to the discovery of two VMEs, mainly composed of calcified organisms, at a depth near to the aragonite saturation horizon, making these VMEs into laboratories to analyze the effects of ocean acidification on benthic communities in the Southern Ocean.

Integrated oceanographic campaigns such as the CEAMARC expedition have become the norm rather than the exception, and the benefits reaped by this multidisciplinary approach are incomparable. The description of biodiversity in all possible aspects includes the description of the environmental conditions which allow these organisms to live in the area. The analysis of such a vast quantity of data inevitably takes a long time, meaning that research on a project basis is impractical. Thus, the execution of such campaigns must necessarily rely on a community of permanent researchers.

3.5. References

Allcock, A.L., Barratt, I., Eléaume, M., Linse, K., Norman, M.D., Smith, P.J., Steinke, D., Stevens, D.W., Strugnell, J.M. (2011). Cryptic speciation and the circumpolarity debate: A case study on endemic Southern Ocean octopuses using the COI barcode of life. *Deep Sea Research Part II: Topical Studies in Oceanography*, 58(1/2), 242–249.

Arango, C.P., Soler-Membrives, A., Miller, K.J. (2011). Genetic differentiation in the circum – Antarctic sea spider Nymphon australe (Pycnogonida; Nymphonidae). *Deep Sea Research Part II: Topical Studies in Oceanography*, 58(1/2), 212–219.

Arnaud, P.M. (1974). Contribution à la bionomie marine benthique des régions antarctiques et subantarctiques. *Thétys*, 6(3), 465–656.

Ausubel, J.H., Trew Crist, D., Waggoner, P.E. (2010). *First Census of Marine Life 2010: Highlights of a Decade of Discovery*. Census of Marine Life, Washington, DC [Online]. Available at: http://www.coml.org/pressreleases/census2010/PDF/Highlights-2010-Report-Low-Res.pdf.

Baird, H.P., Miller, K.J., Stark, J.S. (2011). Evidence of hidden biodiversity, ongoing speciation and diverse patterns of genetic structure in giant Antarctic amphipods: Genetic diversity in Antarctic amphipods. *Molecular Ecology*, 20(16), 3439–3454.

Beaman, R.J. and Harris, P.T. (2003). Seafloor morphology and acoustic facies of the George V Land shelf. *Deep Sea Research Part II: Topical Studies in Oceanography*, 50(8/9), 1343–1355.

Beaman, R.J. and Harris, P.T. (2005). Bioregionalization of the George V Shelf, East Antarctica. *Continental Shelf Research*, 25(14), 1657–1691.

Beaman, R.J. and O'Brien, P.E. (2009). Collaborative east Antarctic marine census (CEAMARC): RSV Aurora Australis Voyage 3, December 2007–January 2008. Post-survey report, Geoscience Australia, Canberra.

Beaman, R.J., O'Brien, P.E., Post, A.L., De Santis, L. (2011). A new high-resolution bathymetry model for the Terre Adélie and George V continental margin, East Antarctica. *Antarctic Science*, 23(1), 95–103.

Brooks, C.M., Chown, S.L., Douglass, L.L., Raymond, B.P., Shaw, J.D., Sylvester, Z.T., Torrens, C.L. (2020). Progress towards a representative network of Southern Ocean protected areas. *PLOS ONE*, 15(4), e0231361.

Causse, R., Ozouf-Costaz, C., Koubbi, P., Lamy, D., Eléaume, M., Dettaï, A., Duhamel, G., Busson, F., Pruvost, P., Post, A. et al. (2011). Demersal ichthyofaunal shelf communities from the Dumont d'Urville Sea (East Antarctica). *Polar Science*, 5(2), 272–285.

Chenuil, A., Saucède, T., Hemery, L.G., Eléaume, M., Féral, J.-P., Améziane, N., David, B., Lecointre, G., Havermans, C. (2018). Understanding processes at the origin of species flocks with a focus on the marine Antarctic fauna: Understanding the origins of species flocks. *Biological Reviews*, 93(1), 481–504.

Dettaï, A., Adamowizc, S.J., Allcock, L., Arango, C.P., Barnes, D.K.A., Barratt, I., Chenuil, A., Couloux, A., Cruaud, C., David, B. et al. (2011). DNA barcoding and molecular systematics of the benthic and demersal organisms of the CEAMARC survey. *Polar Science*, 5(2), 298–312.

Fogg, G.E. (1992). *A History of Antarctic Science*. Cambridge University Press, Cambridge.

Galton-Fenzi, B.K. (2009). Modelling ice-shelf/ocean interaction. PhD Thesis, University of Tasmania, Hobart.

Grant, S., Constable, A.J., Raymond, B., Douste, S. (2006). Bioregionalisation of the southern ocean: Report of the Experts Workshop (Hobart, September 2006). Report, WWF – Australia Head Office, Sydney.

Gutt, J., Koubbi, P., Eléaume, M. (2007). Mega-epibenthic diversity off Terre Adélie (Antarctica) in relation to disturbance. *Polar Biology*, 30, 1323–1329.

Hemery, L.G., Galton-Fenzi, B., Améziane, N., Riddle, M., Rintoul, S., Beaman, R., Post, A., Eléaume, M. (2011). Predicting habitat preferences for *Anthometrina adriani* (Echinodermata) on the East Antarctic continental shelf. *Marine Ecology Progress Series*, 441, 105–116.

Hemery, L.G., Eléaume, M., Roussel, V., Améziane, N., Gallut, C., Steinke, D., Cruaud, C., Couloux, A., Wilson, N.G. (2012). Comprehensive sampling reveals circumpolarity and sympatry in seven mitochondrial lineages of the Southern Ocean crinoid species *Promachocrinus kerguelensis* (Echinodermata): Circumpolarity of seven COI crinoid lineages. *Molecular Ecology*, 21(10), 2502–2518.

Hosie, G., Koubbi, P., Riddle, M., Ozouf-Costaz, C., Moteki, M., Fukuchi, M., Ameziane, N., Ishimaru, T., Goffart, A. (2011). CEAMARC, the collaborative East Antarctic marine census for the census of Antarctic marine life (IPY# 53): An overview. *Polar Science*, 5(2), 75–87.

Hureau, J.C. (1962). Poissons antarctiques récoltés au cours de la onzième expédition française en Terre Adélie 1960–1962. *Bulletin du Muséum national d'histoire naturelle*, 34(2/3), 228–238.

Hureau, J.C. and Arnaud, P. (1964). Complément à la faune de poissons de la Terre Adélie (Antarctique). *Bulletin de l'Institut océanographique de Monaco*, 62(1295), 1–14.

Koubbi, P., Ozouf-Costaz, C., Goarant, A., Moteki, M., Hulley, P.-A., Causse, R., Dettaï, A., Duhamel, G., Pruvost, P., Tavernier, E. (2010). Estimating the biodiversity of the East Antarctic shelf and oceanic zone for ecoregionalisation: Example of the ichthyofauna of the CEAMARC (Collaborative East Antarctic Marine Census) CAML surveys. *Polar Science*, 4(2), 115–133.

Lecointre, G., Améziane, N., Boisselier, M.C., Bonillo, C., Busson, F., Causse, R., Chenuil, A., Couloux, A., Coutanceau, J.-P., Cruaud, C. et al. (2013). Is the species flock concept operational? The Antarctic shelf case. *PloS one*, 8(8), e68787.

Post, A.L., Beaman, R.J., O'Brien, P.E., Eléaume, M., Riddle, M.J. (2011). Community structure and benthic habitats across the George V Shelf, East Antarctica: Trends through space and time. *Deep Sea Research Part II: Topical Studies in Oceanography*, 58(1/2), 105–118.

Rosenberg, M. and Rintoul, S.R. (2010). Southern Ocean CTD data collected from the CEAMARC cruise of the Aurora Australis, 2007–2008. Australia Ocean Data Network [Online]. Available at: http://data.aad.gov.au/metadata/CEAMARC_CASO_200708_V3_CTD [Accessed 24 May 2023].

Soler-Membrives, A., Linse, K., Miller, K.J., Arango, C.P. (2017). Genetic signature of last glacial maximum regional refugia in a circum-Antarctic sea spider. *Royal Society Open Science*, 4(10), 170615.

4

Objectif Plancton: A Citizen Science Program to Study Plankton Diversity

4.1. A citizen science project

Objectif Plancton is a scientific research program driven by a citizen initiative. This regional project brings together scientists, pleasure boaters, scientific mediators and the general public, to further their knowledge of the ocean at three sites on the coast of Brittany: Brest and Lorient roadsteads and Concarneau Bay.

The unique aspect of Objectif Plancton is, depending on the site, to gather between 5 and 17 pleasure boats that simultaneously collect samples at different points in the study area, providing a synoptic view. This represents an unprecedented opportunity for researchers, who do not have the resources to carry out this type of sampling for themselves. This operation takes place three times a year during the same period: at Brest, Lorient and Concarneau, in April, June and September. There are peculiarities specific to each site, but overall, the way in which a session is run is similar from one site to another. The sessions take place on Saturdays in Brest and Concarneau, and on Wednesdays in Lorient. In the morning, the mediators hand out a kit containing the material to each participating boat, giving all of the necessary

Chapter written by Cyril GALLUT, Céline LIRET, Marine LE MOAL, Philippe PONDAVEN, Cécile BANOVSKI-KLEIN, Mathilde CADIER, Antoine CHARPENTIER, Klervi FUSTEC, Julianne LE GUEN, Molène LE ROY, Joëlle PICHON, Martin PLUS, Laura SCHWEIBOLD, Marc SOURISSEAU and Nalani K. SCHNELL.

explanations regarding sampling protocols. In addition to pleasure boaters, others can participate if there are spaces available (often, these others are students or nature lovers). The boats sail to the GPS point assigned to them and take the required measurements and samples at the set time. They then bring the samples back to shore, where the mediators and scientists prepare them and preserve them so they can be analyzed in the lab. This is a key time for participants, mediators and scientists to meet and exchange, and also offers the opportunity for everyone to discover the fascinating diversity of plankton, and to learn how fragile that type of biodiversity is. Once a year, at each of the sites, the participants and the general public are invited to an evening event to hear about the advancement of the project.

It was in 2014 that Océanopolis first organized plankton collection operations by pleasure boat users in the roadsteads at Brest and Lorient, with a view to raising public awareness of the importance of plankton in the ocean. This civic participation represented a unique opportunity to simultaneously sample different points in the same ecosystem. The scientists seized the opportunity to study the spatial distribution of plankton on a small scale.

This event gave rise to the Objectif Plancton program, which gradually took shape. At each site, a scientific mediation body was entrusted with bringing pleasure boaters on board at a local level, facilitating the link between scientists and seagoers, developing knowledge distribution and sharing with the participants and with the public at large: Océanopolis in Brest; the Observatoire du Plancton (Plankton Observatory) in Lorient; and the Explore fund in Concarneau. Various scientific partners are involved in the project: the Institut universitaire européen de la mer (IUEM-UBO – (European Institute for Marine Studies)), the Institut français de recherche pour l'exploitation de la mer (IFRMER (French Marine Exploitation Research Institute)) and the Station marine de Concarneau (Muséum national d'Histoire naturelle).

In terms of seagoers, multiple bodies are involved as well: the Amicale des plaisanciers des marinas de Brest (APMB (Brest Marinas Pleasure Boaters Association)); the Brest SNSM; the Centre nautique de Fouesnant Cornouaille; the Club nautique des minahouets de Locmiquélic; the Association des pêcheurs plaisanciers de la région de Lorient; the Amicale des pêcheurs plaisanciers de Port-Louis; as well as the associations Les Glénans, Marche Avec, Ansel and Astrolabe Expéditions. Océanopolis handles the overall coordination of these participants (see Figure 4.1).

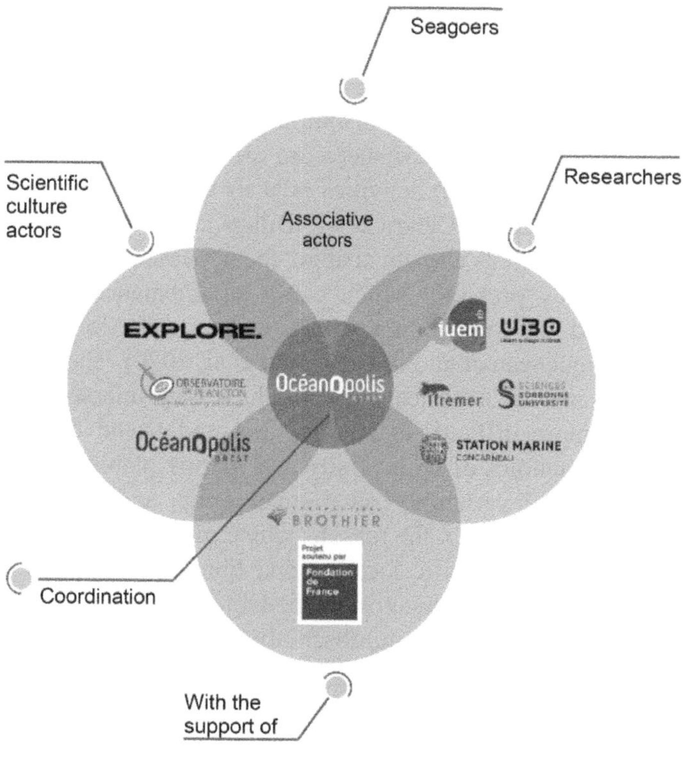

Figure 4.1. *The Objectif Plancton "ecosystem". For a color version of this figure, see www.iste.co.uk/nicolas/biodiversity.zip*

4.2. Scientific objectives

The ocean is home to a multitude of organisms which float and drift with the currents. This life is known as plankton, and is typically invisible to the naked eye. It represents over 95% of marine biomass and is made up of astonishingly diverse lifeforms: viruses, bacteria, microscopic algae, reproductive cells, fish larvae, etc. The particle size ranges from 0.1 μm for bacteria and viruses up to over a meter for certain jellyfish. Plankton plays an absolutely crucial role in ocean functioning, and in the balance of our planet. Phytoplankton produces more than 50% of the O_2 in the air that we breathe, and helps regulate the climate. It is also at the base of practically all marine food chains.

At present, all of the factors which influence plankton biology and ecology (temperature, nutrient content, pH, redox state and availability of light) are directly impacted by human activities (Hutchins and Fu 2017), which has consequences for the geographic distribution of species, biodiversity and the dynamic of plankton communities (for example, see Oziel et al. 2020). These disturbances are all the greater in coastal areas, where pressure due to human activities is at its strongest. In addition to the process of eutrophication (Cloern 2001; Billen and Garnier 2007), we see other disturbances such as climate change, the appearance of invasive exotic species, and fishing pressure, which cause an imbalance in the plankton communities and the functioning of the ecosystems (Reid et al. 2000; Cloern 2001; Ragueneau et al. 2018; Bestion et al. 2020). The multiplicity of these pressures and the nonlinear way in which they interact (Hewitt et al. 2016) make it more difficult to identify and understand the mechanisms controlling the dynamics of biodiversity and marine plankton communities. Assessing the spatial variability on a small scale, and understanding the underlying mechanisms – i.e. the reasons why a given species, as opposed to another, develops at a specific location and time of year – is therefore of major interest. The specific objective of the Objectif Plancton program, then, is to conduct (and connect) in-situ observations and studies of processes, at different spatial and temporal scales, to improve our understanding of the mechanisms controlling the dynamics of coastal plankton communities.

Multiple topics are addressed: the diversity, structure and dynamics of the coastal plankton communities (spatial, seasonal and multiannual); inventory of the diversity of coastal ichthyoplankton (fish eggs and larvae) and their relation to plankton composition; detection and monitoring of toxic microalgae; and integration of the causes and risks that may be associated with them.

The establishment of Objectif Plancton in 2014 led to the emergence of new questions, on top of the initial research questions. Firstly, the plankton analyses focused on studying phytoplankton and ichthyoplankton. Secondly, Objectif Plancton is a citizen science program, relying on the involvement of pleasure boaters and other lay participants, mediators and scientists. Each of the actors involved has varied and multiple relationships with the others.

They have expressed the need to better understand everyone's respective roles and the interactions among the different stakeholders, to enable the program to become part of regional activities over the long term. Thus, a new facet – of human and social science – has been part of the Objectif Plancton program since 2022.

Citizen science covers a range of scientific fields, such as the natural sciences, medical sciences, engineering, human and social sciences (Haklay et al. 2021) and a wide range of topics. In recent years, citizen science projects relating to marine and coastal problems have been developing (Thiel et al. 2014) but little space is given to these issues in comparison with land areas (Lagourgue 2015; Garcia-Soto et al. 2017; Sauleau et al. 2021). The sociological study takes an in-depth look at a citizen science program (Objectif Plancton), to contribute to better knowledge of the links between science and society in coastal and marine environments.

The objective is to understand the way in which researchers in human and social sciences and calls for projects define citizen science, and to examine these processes with an in-depth study of the Objectif Plancton program in the field. It is a matter of focusing on the actors who facilitate the production of knowledge about plankton. To do this, we combine three methods of inquiry: a review of the scientific and so-called gray literature concerning the process of definition of citizen sciences; semi-structured interviews with the pleasure boaters, scientists, mediators and coordinators; and the establishment of a participant observation system for use while the samples are being collected, at workshops where sampling gear is being made or repaired, and at meetings of the various internal committees that are part of the program.

4.3. Materials and methodology

Objectif Plancton events take place in the roadsteads of Brest and Lorient, and in the Concarneau Bay (see Figure 4.2). In Brest, the study area is covered by 17 sampling points; there are 7 points in Lorient roadstead; and 5 in the Concarneau Bay (see Figure 4.3). Thus, there are 29 points in total. The sampling sessions take place three times a year at each site, in April, June and September.

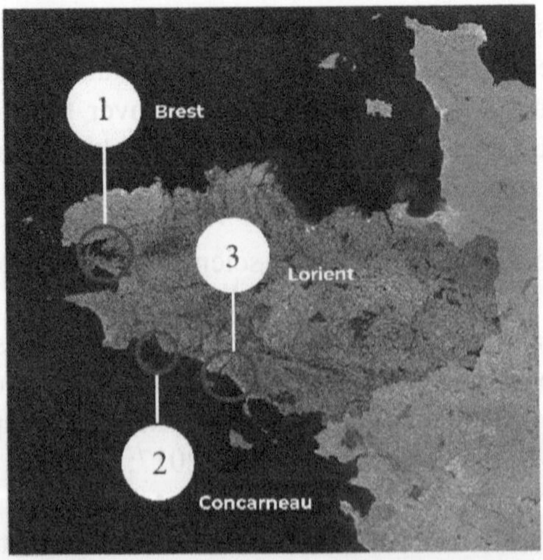

Figure 4.2. *The three sites studied by Objectif Plancton. For a color version of this figure, see www.iste.co.uk/nicolas/biodiversity.zip*

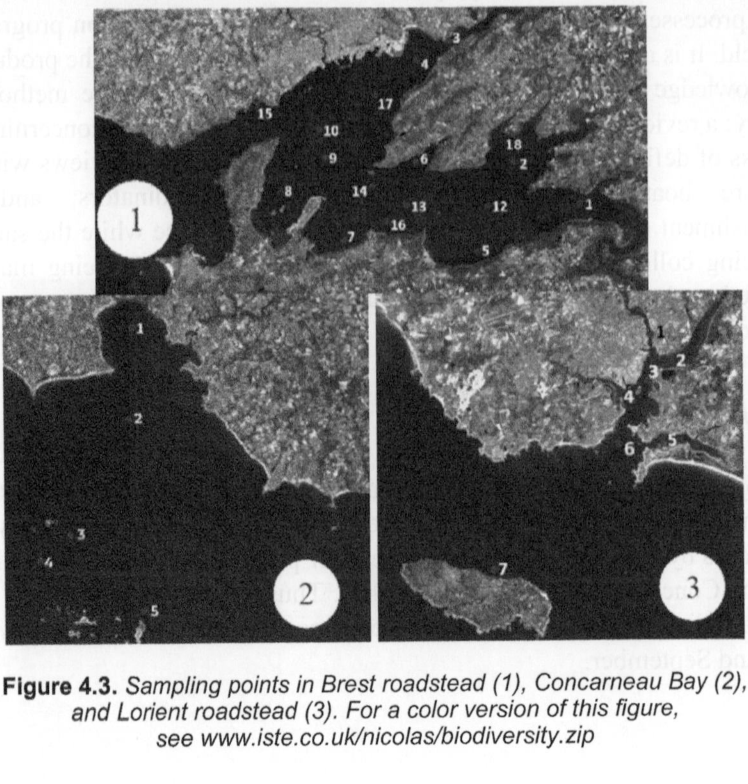

Figure 4.3. *Sampling points in Brest roadstead (1), Concarneau Bay (2), and Lorient roadstead (3). For a color version of this figure, see www.iste.co.uk/nicolas/biodiversity.zip*

At the outset of Objectif Plancton, the scientists, mediators and also the participants in the program worked together on the design and adaptation of collection tools that would be easy for pleasure boaters to use. Indeed, the scientific gear that is used aboard oceanography vessels is too technical, too costly and too bulky for the small craft – such as sailboats and semi-rigid boats – which make up the Objectif Plancton "fleet". Workshops were organized for several years running to make ichthyoplankton nets (serving to sample teleostean fish eggs and larvae), to equip the three sites. These nets must inevitably be "homemade", because there are currently no nets on the market in the configuration used for the purposes of Objectif Plancton. The scientists and mediators also worked on collection protocols, so they had a document, shared across the three sites, detailing the various types of samples that were to be collected. These protocols guide the participants through the procedures for collecting samples at sea, and how to store and protect those samples before bringing them back to shore. This document contains the same information as the explanations and demonstrations given by the scientific mediators prior to each excursion on the sea. The same instructions are summarized in a video which participants can view aboard the vessels, in case of doubt.

A considerable amount of work has also been invested in ensuring the sampling kits available are consistent across the three sites. These kits comprise the following tools:

– a Secchi disk to measure water turbidity;

– various containers (jars and bottles) for sample storage;

– a thermometer to measure the water temperature at the sampling point;

– an isothermal bag and ice to preserve certain samples (phytoplankton and ichthyoplankton);

– different types of nets depending on what needs to be sampled for lab analysis (phytoplankton net 20 µm; ichthyoplankton net 500 µm, 50 cm diameter mouth opening);

– a collector tube to collect seawater for the harvesting of phytoplankton and analysis, in the lab, of the following parameters: salinity, nutrient concentrations, photosynthetic pigment levels, cell abundance, total biomass and make-up of organic matter;

– acid Lugol's solution to preserve certain samples of phytoplankton until they can be analyzed under optical microscopy in the lab;

– a data sheet to record the associated metadata.

In order to analyze and compare the sites over the long term, a set of parameters are recorded on each outing at each of the sampling points:

– temperature: measuring the seawater temperature to discover spatial variations which could account for the different habitats and the changes over several years;

– pH: measuring the hydrogen potential of the water (that is, its concentration of hydrogen ions), whereby its acidity or basicity can be defined;

– salinity: measuring the quantity of dissolved salts in the water. This value can be found by measuring conductivity;

– turbidity: estimating the amount of suspended matter by measuring water clarity;

– chlorophyl *a*: measuring that pigment, which allows algae to capture the light necessary for photosynthesis, is a way of determining the biomass of phytoplankton;

– nutrient salts: quantifying the concentrations of nitrates and nitrites, phosphates and silicates present in the seawater. These major nutrients can help to link plankton diversity and distribution to certain environmental factors;

– cell abundance of photosynthetic organisms: by cytometry, all cellular organisms (even those which are tiny, and invisible with an optical microscope) are counted and classified on the basis of their overall pigmentary composition;

– phytoplankton taxonomy: describing and listing the diversity of plant-based plankton under an optical microscope;

– ichthyoplankton taxonomy: describing the diversity of fish larvae and eggs.

The participants, returning to shore, bring back the sampling equipment, the samples collected, and the sheet containing the associated metadata. The samples are processed by the scientists and mediators to condition them for preservation and subsequent analysis in the laboratory.

4.4. Avenues of research

Two axes structure the research work of the Objectif Plancton program: one focusing on zooplankton (particularly, fish larvae), and one on phytoplankton. Within this latter axis, there are three subdivisions: an environmental aspect, with analysis of nutrients and their link to plant-based plankton (in Brest and Lorient roadsteads); a focus on determining diversity (taxonomy and cytometry); and a focus on toxic algae (in Brest roadstead).

4.4.1. *Ichthyoplankton*

The last research on ichthyoplankton in this area dated from the late 1960s. Thus, it appeared necessary to update these old data. The aim is to study the diversity of fish larvae present in this area. The species are inventoried and described, both morphologically and genetically. Thus, Objectif Plancton enables scientists to develop spatiotemporal monitoring of these fish larvae, and eventually, to analyze the fluctuations of this zooplankton in relation to global change. Identification is carried out by two different methods: on the basis of observations with a stereomicroscope (ZEISS© SteREO Discovery.V20) with a camera lucida (ZEISS© Axiocam 506 color), and with genetic identification if necessary. The fish larvae and eggs (see Figure 4.4) are identified, as far as possible, to species level, on the basis of morphological characteristics.

The characteristics used for identification of the larvae are as follows: general shape, length of the intestine, arrangement of pigmentation (melanophores) and the number of myomeres (Russell 1976). Certain species, in the same family or the same genus, are very difficult to distinguish from one another on the basis of morphological criteria. In such cases, the specimens are also analyzed by genetic identification. A tissue sample is taken from the fish larva to extract its genetic material; then the genetic sequence of a reference fragment (a portion of the gene coding for cytochrome c oxidase subunit I) is characterized, and serves as a genetic barcode. The genetic sequence obtained is compared to a reference library containing millions of sequences from numerous species (Ratnasingham and Hebert 2007). The unknown larva can thus be identified. Once identified, the specimens are inventoried and added to the collection of the Muséum national d'Histoire naturelle. Those are the first samples of larval fish from the Northeast Atlantic deposited in the MNHN collection. Between 2015 and 2022, around 1300 larvae were collected. These larvae correspond to 34 species belonging to 20 different families. The project will continue in years to come, and on the basis of these data, any future changes in the species composition can be detected and viewed in relation to the surrounding environmental parameters.

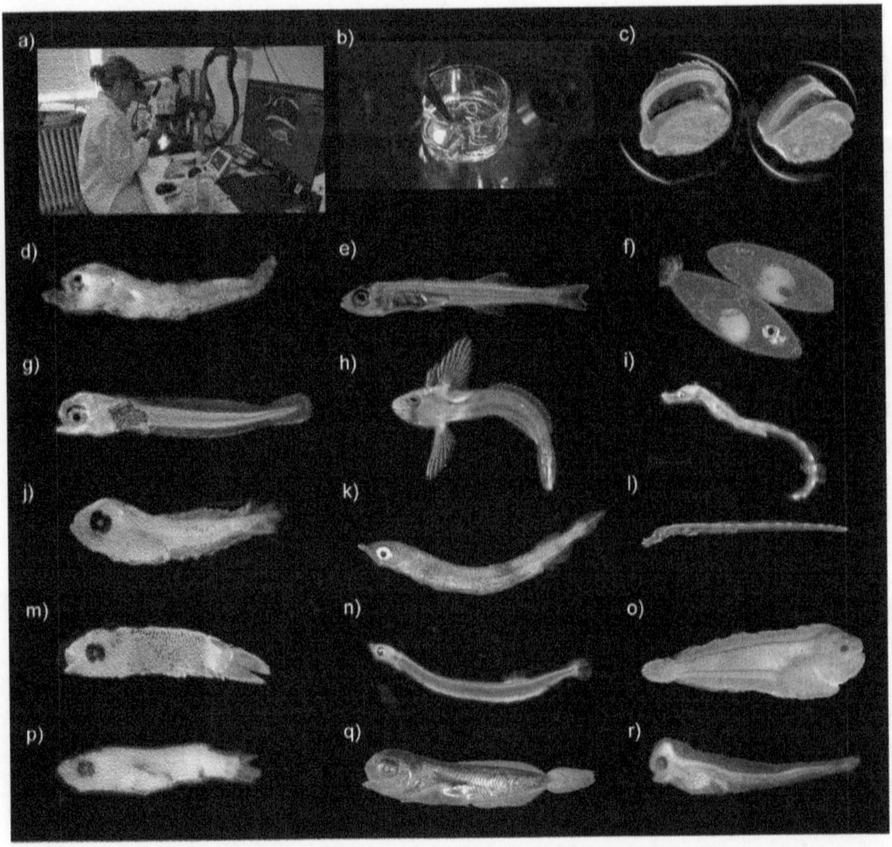

Figure 4.4. *Morphological identification of fish larvae and a few examples of larvae. For a color version of this figure, see www.iste.co.uk/nicolas/biodiversity.zip*

COMMENT ON FIGURE 4.4.– *a) Observations with the stereomicroscope (ZEISS© SteREO Discovery.V20) with camera lucida (ZEISS© Axiocam 506 color) © Pierre-François Watras INTERNEP; b) sample © Pierre-François Watras*

*INTERNEP; c) sardine eggs (*Sardina pilchardus*); d) goby (*Gobiidae sp.*); e) sand smelt (*Atherina presbyter*); f) goby eggs (*Gobiidae sp.*); g, h) shanny (*Lipophrys pholis*); i) hippocampus (*Hippocampus sp.*); j) black seabream (*Spondyliosoma cantharus*); k) garfish (*Belone belone*); l) syngnathid (*Syngnathidae sp.*); m) corkwing wrasse (*Symphodus melops*); n) sardine (*Sardina pilchardus*); o) common sole (*Solea solea*); p) goldsinny wrasse (*Ctenolabrus rupestris*); q) rockling (*Gaidropsaridae sp.*); r) brill (*Scophthalmus rhombus*).*

4.4.2. *Phytoplankton*

Coupled analysis of the spatiotemporal variations in phytoplankton communities and the environmental data was, up until 2020, performed only for Brest roadstead. Now, these analyses are conducted at all three sites. Samples of phytoplankton taken in September 2014, September 2019 and September 2020 in Brest roadstead have been analyzed by optical microscopy, and coupled with various environmental parameters (salinity, temperature and concentrations of nutrient salts). This study demonstrates spatial variability (at the scale of a few kilometers) and temporal variability of phytoplankton communities (> 5 µm) sampled at the same time of year. In September 2014, the dominant species in the roadstead were nanophytoplankton microalgae (diameter < 10 µm, see Figure 4.5), which accounted for 85% of the total abundance for each station, whereas at the same point in 2020, 90% of specimens sampled were microphytoplankton diatoms (> 10 µm, see Figure 4.6). With respect to spatial variability, the composition of the phytoplankton communities within the roadstead is not homogeneous, with zones that may display local accumulations of such-and-such a taxon. From a spatial point of view, these local accumulations, which have been correlated with the presence of certain nutrient salts, and the increase or decrease in salinity and temperature, are dependent on the circulation of water masses and inflows from the land. They may have implications, notably in terms of monitoring and understanding the dynamics of blooms of toxic species (for example, the dinoflagellate *Alexandrium minutum*). Continuation of such spatial monitoring, as part of the Objectif Plancton program, should in time help to refine our knowledge of the phenology of phytoplankton blooms in Brest roadstead.

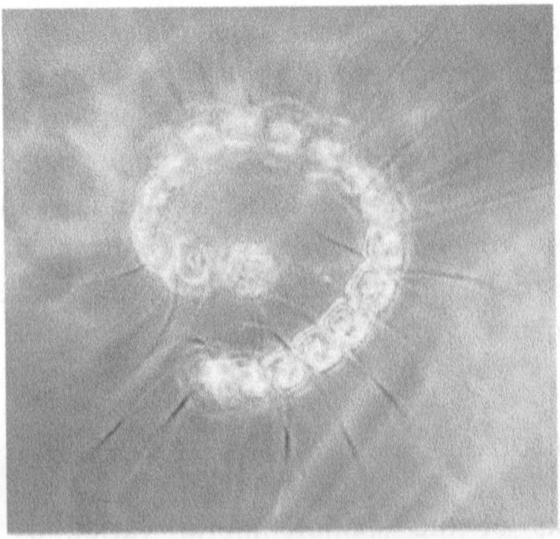

Figure 4.5. *Nanophytoplankton: chain of* Chaetoceros cf debilis, *(10 µm per cell) (source: Laura Schweibold). For a color version of this figure, see www.iste.co.uk/nicolas/biodiversity.zip*

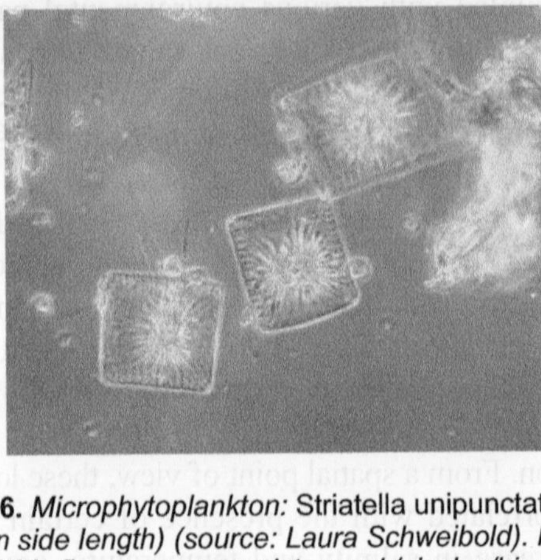

Figure 4.6. *Microphytoplankton:* Striatella unipunctata, *(around 100 µm in side length) (source: Laura Schweibold). For a color version of this figure, see www.iste.co.uk/nicolas/biodiversity.zip*

Multiple methods are used for analyzing the samples of phytoplankton: chemotaxonomic analyses, FlowCam, cytometry and microscopy. Chemotaxonomic analyses, using the analytical tool Chemtax, are capable of identifying the main groups of plant-based plankton present in each sample, thanks to the signature of their photosynthetic pigments, identified and quantified for each sample using HPLC. Thus, it is possible to determine the relative contributions of phytoplanktonic groups with respect to the total abundance contained in a sample. In parallel, the samples are analyzed using the FlowCam, the principle of which is simple: a flow of water from the collected samples is sent through the machine, which photographs and measures each particle present, providing numerous categories of associated data (size, diameter, biovolume, concentration, fluorescence, etc.). This method has the advantage of creating an image bank of the communities present at time t in the roadstead over the long term. Cytometric analysis determines the total abundance of phytoplanktonic cells across the whole size spectrum (from 0.2 μm to several millimeters). They are determined in broad groups on the basis of their fluorescence values which, themselves, are dependent on their pigmentary compositions. The results exhibit significant and rapid variations in the spatial distribution of these groups in Brest roadstead, both seasonally and year on year. Analyses using microscopy take a great deal of time, but they have the advantage of being reliable and accurate. These tools can be used to work on the taxonomy of phytoplankton in Brest roadstead, and to count and describe the phytoplanktonic organisms captured through sampling. The class, genus and species of the dominant taxa are determined for each sample in order to calculate diversity indices and compare the proportions of the different classes of phytoplankton between the sampling points, firstly, and across seasons and years, secondly. Thus, it has been shown that there is strong spatial structuring of the phytoplankton communities across Brest roadstead.

The samples collected as part of the Objectif Plancton program are of interest to certain scientists specializing in other disciplines. For instance, researchers studying microplastics made a request of the Objectif Plancton team to obtain samples in Brest roadstead, in order to discover the abundance of particles of microplastics and their distribution, with a view to studying the dynamic of circulation of these particles within the roadstead.

4.4.3. *Human and social sciences*

The program Objectif Plancton focuses on the study of plankton. Analysis in human and social sciences focuses on the human beings who facilitate knowledge production about plankton. The program involves a range of actors whom we have grouped together as follows: scientists, scientific mediators, participants and coordinators. Each type of actor participates in the project for various reasons, and they have different roles to play within the program. For analytical purposes, we have created ideotypes: categories which can be used for analysis, which provide clues to the understanding of the actors' motives for contributing to a citizen science program, but the list is by no means exhaustive.

What are the interests which lead researchers to engage in the citizen science program, Objectif Plancton? The most obvious point is the contribution of material and logistical support. Scientists do not have enough boats to implement a protocol such as that described above. Next, researchers participate with the desire to transmit knowledge; a desire that is also found in the other actors involved. Within the program, the scientists work with interns (from undergraduate to postgraduate level), and supervise doctoral candidates. During the days of collection and delivery of results, they give a presentation on what their profession involves, and of the scientific questions linked to plankton. Finally, the researchers take pleasure in carrying out their jobs in this context, in the field. They have conversations with people whom they would not normally meet in their day-to-day lives, and answer questions that are different to those posed by their academic peers.

What is the role of the scientific mediators? The idea of transmission is crucially important. The mediators transmit a range of types of knowledge: technical knowledge (how to handle the equipment while at sea), knowledge about plankton, the sciences, the profession of research, and environmental issues (plastic pollution, for example). They also help us to convey information and questions submitted to them by the participants to the scientists (a collector tube that is a little tricky to handle, for example). They organize day events for collection and delivery of results, often in collaboration with the scientists and coordinators. They are in charge of contacting participants, finding the correct number of boats, putting different people together to form the necessary crews, preparing the equipment, etc. They help the events to run smoothly, to ensure that the samples are collected and submitted to the scientists in accordance with the protocol.

What motivations do participants have for joining in a program such as Objectif Plancton? Participants speak of the joy they feel at contributing to the program: the pleasure of a trip out to sea, of welcoming people on their boats, of coming together, contributing to a scientific project, looking at the plankton, etc. Curiosity is also one of the driving forces behind participation (the same is true for other categories of actors as well). They are very eager to discover the plankton: what is there in the samples that they have brought back? They learn to see plankton with the naked eye in the samples collected, but also to identify certain species of plankton once the sample is under the microscope. Finally, knowledge transfer is also a key element in the decision to participate. This knowledge takes a variety of forms: knowledge about science and plankton; knowledge about the landscape and the surrounding environment; knowledge about the practice of pleasure-boating, which is transmitted by inviting schoolchildren, students and friends aboard the boats; the coming of family and children, especially; the forging of new relationships, etc.

In what way do the coordinators contribute to the program? They ensure that it runs smoothly. They initiate the hunt for funding to allow the program to continue in the long term (taking on interns and doctoral candidates, producing and maintaining collection equipment, producing mediation tools, etc.). The coordinators also handle the promotion and valorization of the program. Alongside the other actors, they present the program in a variety of contexts: ocean days, discussion days involving the stakeholders, etc. The coordinators take care of the structuring of the program and the linkage between the sites (by running inter-site collection days, where the actors from one site participate in a collection at the other sites, once a year, for example), and linkage among the actors (by organizing scientific committees, mediation and communication). The coordinators are the point of contact for all other actors, so they act as a facilitator.

Thus, while the roles and motivation of the various people involved may be broadly divided into groups of actors, the reality in the field is different: often, roles blend into one another, and the motives are shared.

4.5. Conclusion

The Objectif Plancton program, which arose from the coming together of pleasure boaters, scientific mediators and researchers, is a regional citizen science program, which aims to look at the long term. Its primary objective is to study the mechanisms underlying the temporal and spatial variability of plankton in a temperate coastal environment, with a large tidal range. The participation of numerous pleasure boaters in sessions to collect samples provides an unprecedented view of the composition of the plankton communities. The initial results have helped them to compile an inventory of teleostean fish larvae from the study area, to analyze the composition and dynamics of the phytoplankton communities and toxic phytoplankton species. Objectif Plancton also represents a wonderful opportunity to raise public awareness of the extraordinary richness of plankton, its role in the operation of marine ecosystems, and the absolute necessity of protecting it.

Over recent years, the structuring of the Objectif Plancton program has helped forge close links among the various actors involved. This collective building of the link between science and society has led to the identification of new requirements, specific and common expectations, and the willingness to share through new actions. In particular, it has proven important to study the whole of the zooplanktonic compartment, which will therefore be the subject of future developments in the project. Finally, on the strength of its success in Europe, Objectif Plancton has been expanded to Quebec, where a new site is to be set up in Saguenay Fjord.

4.6. References

Bestion, E., Barton, S., García, F.C., Warfield, R., Yvon-Durocher, G. (2020). Abrupt declines in marine phytoplankton production driven by warming and biodiversity loss in a microcosm experiment. *Ecol. Lett.*, 23(3), 457–466.

Billen, G. and Garnier, J. (2007). River basin nutrient delivery to the coastal sea: Assessing its potential to sustain new production of non-siliceous algae. *Mar. Chem.*, 106(1/2), 148–160.

Cloern, J.E. (2001). Our evolving conceptual model of the coastal eutrophication problem. *Mar. Ecol. Prog. Ser.*, 210, 223–253.

Garcia-Soto, C., Van der Meeren, G.I., Busch, J.A., Delany, J., Domegan, C., Dubsky, K., Fauville, G., Gorsky, G. (2017). *Advancing Citizen Science for Coastal and Ocean Research*. European Marine Board, Ostend.

Haklay, M., Fraisl, D., Greshake Tzovaras, B., Hecker, S., Gold, M., Hager, G., Ceccaroni, L., Kieslinger, B., Wehn, U., Woods, S. et al. (2021). Contours of citizen science: A vignette study. *R. Soc. Open Sci.*, 8(8), 202108.

Hewitt, J.E., Ellis, J.I., Trush, S.F. (2016). Multiple stressors, nonlinear effects and the implications of climate change impacts on marine coastal ecosystems. *Glob. Chang. Biol.*, 22(8), 2665–2675.

Hutchins, D.A. and Fu, F. (2017). Microorganisms and ocean global change. *Nat. Microbiol.*, 2, 17058.

Lagourgue, L. (2015). État des lieux de la science participative sur les algues et proposition de voies d'utilisation et de valorisation des données intrantes. PhD Thesis, Université de Sherbrooke, Sherbrooke.

Oziel, L., Baudena, A., Ardyna, M., Massicotte, P., Randelhoff, A., Sallée, J.-B., Ingvaldsen, R.B., Devred, E., Babin, M. (2020). Faster Atlantic currents drive poleward expansion of temperate phytoplankton in the Arctic Ocean. *Nat. Commun.*, 11, 1705.

Ragueneau, O., Raimonet, M., Mazé, C., Coston-Guarini, J., Chauvaud, L., Danto, A., Grall, J., Jean, F., Paulet, Y.-M., Thouzeau, G. (2018). The impossible sustainability of the Bay of Brest? Fifty years of ecosystem changes, interdisciplinary knowledge construction and key questions at the science-policy-community interface. *Front. Mar. Sci.*, 5, 124.

Ratnasingham, S. and Hebert, P.D.N. (2007). Bold: The barcode of life data system (http://www.barcodinglife.org). *Mol. Ecol. Notes*, 7(3), 355–364.

Reid, P.C., Battle, E.J.V., Batten, S.D., Brander, K.M. (2000). Impact of fisheries on plankton community structure. *ICES J. Mar. Sci.*, 57(3), 495–502.

Russell, F.S. (1976). *The Eggs and Planktonic Stages of British Marine Fishes*. Academic Press, London.

Sauleau, P., Poisson, P., Ruault, R. (2021). Les sciences participatives dédiées à la mer et au littoral breton : état des lieux, recommandations et prospectives pour les universités ISblue. Report, IUEM Institut Européen de la Mer, Plouzané.

Thiel, M., Penna-Díaz, M.A., Luna-Jorquera, G., Salas, S., Sellanes, J., Stotz, W. (2014). Citizen scientists and marine research: Volunteer participants, their contributions, and projection for the future. In *Oceanography and Marine Biology: An Annual Review*, Gibson, R.N., Atkinson, R.J.A., Gordon, J.D.M. (eds). CRC Press, Boca Raton.

5

Environmental DNA for Observing Marine Mammals in the Marine Protected Areas of Iroise and the Antilles

5.1. Introduction

The astonishing breaches of humpback whales, dolphins frolicking around boats, the sheer majesty of pods of orcas, the breathtaking size of blue whales, and the dreamy gentleness of sperm whales… these sights leave a lasting impression on anyone lucky enough to see them. It has become a widely held misconception that we (the public at large, and the scientific community) know all that there is to know about whales, dolphins, porpoises and other orcas. These animals are iconic, plainly visible, have been studied for many years, and thus, we would imagine, they are well known.

However, on a regular basis, new species are being identified, such as the new species of *Berardius* (beaked whales), found in the North Pacific (Morin et al. 2016), or the species of whale found in the Gulf of Mexico (Rosel et al. 2021).

For decades, not a year has gone by without scientific research results shedding new light on the evolution of the cetaceans, the complexity of their social organization, their behaviors, and the diversity demonstrated within

Chapter written by Jean-Luc JUNG.

the species (for example, see Foote et al. 2016; Whitehead 2017; Richard et al. 2018).

Whilst the ban on industrial whaling is now largely respected, global change and local human activities have taken over and had a catastrophic impact on the marine environment and the creatures who inhabit it, including cetaceans. There is a great deal of study needed in order to understand the changes in progress in the world's oceans, the results of which will relate to fundamental research and also research applied to marine environmental conservation.

Peppered throughout the world's oceans, France has the second largest exclusive economic area, after the USA. Today, over 30% of that surface area lies in marine protected areas (MPAs), and protection of cetaceans is a major concern, found in the management plans for the MPAs, and taken into account in European directives such as the Marine Strategy Framework Directive (2008/56/EC). For example, the choice of the site for the establishment of the first natural marine park in France, the Iroise Sea, was influenced by the presence, in that zone, of the greatest species diversity of marine mammals in mainland France. Sanctuaries devoted specifically to marine mammals have even been established in the Antilles (the Agoa Sanctuary) and in the Mediterranean (the Pélagos Sanctuary).

It has become apparent that we need to be more familiar with the ecosystems and the species to which they are home, in order to better protect them, and to develop means of acceptable co-evolution striking the balance between maintaining biodiversity and human activity. With respect to cetaceans, and every time for every MPA, we must ask ourselves the following questions:

– Which cetacean species are present in the area?

– Are they present year round, or only for part of the year? Do they stick faithfully to that particular zone? Do they use it for some specific purpose?

– Are the representatives of the species which we observe here the same[1] as those observed elsewhere?

1 This question may arise in relation to individuals – are we dealing with individuals moving from one study area to another? – but also in relation to groups and indeed entire populations: do the individuals from the different zones interact socially with one another? Is mating and reproduction between individuals from two different zones a frequent occurrence, an occasional one, a rare phenomenon, or does it never happen at all?

– Which human activities may pose a threat to these animals, either directly or indirectly?

5.2. Studying cetaceans in order to know them better: from visual observation to DNA analysis

Cetaceans can be studied, for example, by observation from a boat, from the air or, more rarely, from the water itself – citizen science has an important role to play in this type of approach (Coché et al. 2021). In addition, these are animals which make prodigious use of sound to communicate and socialize (Cazau et al. 2016); and in the case of some, the odontoceti, to hunt and to orientate themselves using echolocation. Acoustic monitoring therefore tells us a great deal. Cetaceans can also be studied through their DNA, and many facts that we now understand about them have been discovered through such approaches in recent years (for instance, see Jung 2017; Jung and Madon 2021). Studies in the fields of taxonomy (identification of species, notably cryptic species), systematics (which studies the organization and relations between species and higher taxa), genetic diversity (variations within a species), and social organization (influence of family relationships on the formation of social groups, the transfer of genes between groups), are all examples of understanding given to us by studies on the DNA of cetaceans.

In all cases, though, in order to study DNA, we must have samples. For many years, we have been trying to use what can be described as "non-invasive" samples. For us, the effort began in Madagascar: humpback whales, when breaching as they do so spectacularly, spontaneously slough off skin fragments which remain on the surface of the sea, and are then collected by eco-volunteers from the Madagascan association Cétamada. In Mauritius, François Sarano and his team observe and individually identify sperm whales on dives (Sarano et al. 2022). During these dives, they manually collect fragments of dead skin. In the Iroise Sea, with agents from the natural marine park of Iroise, we use swabs to pass along the backs of large dolphins, thus collecting cells (Jung et al. 2022). All these samples taken non-invasively can be used in the laboratory, with varying degrees of ease (for instance, see Richard et al. 1994; Girardet et al. 2022; Jung et al. 2022).

Of course, sampling just the seawater, and being able to use it, to a certain point, to study the DNA of cetaceans, would be the least invasive source of samples imaginable.

5.3. Progress in approaches using environmental DNA (eDNA) and metabarcoding

Hebert et al. (2003) put forward an approach based on the sequencing of a short fragment (± 500 pb) of the mitochondrial gene coding for subunit 1 of cytochrome c oxidase (CO1) to identify the species from which an animal sample had come. This approach, now having become a standard, we speak of *DNA barcoding stricto sensu* (Valentini et al. 2009), which offers the phenomenal advantage of being able to build up a unified reference database, which can facilitate the work of taxonomists everywhere. Valentini et al. (2009) expanded on the concept, and defined *DNA barcoding* sensu lato, which is the identification of a taxon (e.g. a species, genus, family or even subspecies) by sequencing of a DNA fragment – which may or may not correspond to CO1 – and comparing the sequence to a pre-existing reference database.

Subsequently, next-generation DNA sequencing managed to break down other technological barriers, allowing multiple DNA fragments to be sequenced in parallel, in the same reaction. This opened the door to using metabarcoding to study environmental DNA (eDNA) samples, offering up previously unknown options for environmental monitoring. Environmental DNA is defined as DNA that can be extracted from an environmental sample of earth, water or air, without prior isolation of cells or organisms (Taberlet et al. 2018). Thus, it is a matter of studying a mixture of DNA molecules, in varying states of degradation, coming from a range of organisms whose DNA is present in the sample. The results of this study are a list of species, or of higher or lower taxa, whose DNA is present in the environmental sample. In the marine environment, this entails sampling seawater and, after analysis, listing the organisms whose DNA is present. This method has yielded excellent results in relation to, for example, bony fish and cartilaginous fish (Boussarie et al. 2018; Fernandez et al. 2021; Polanco et al. 2021; Sanchez et al. 2022; Xiong et al. 2022).

The amazing effectiveness of this approach was soon proven. So too, though, were the risks associated with it. There are two types of such risks:

firstly, false positives (a species being identified, even though it is *not* present in the sample – for example, due to experiment contamination in the field or in the lab), and secondly, on the contrary, false negatives (a species failing to be detected, although it *is* present in the sample; this can happen because the detection threshold for the experiment is set too high). The risk of false negatives and false positives stems from the very principle of the approach, requiring us to work with DNA traces which are often badly degraded.

5.4. Detection of marine mammals by eDNA analysis

Following a relatively successful preliminary experiment, targeting harbor porpoises in the Baltic Sea (Foote et al. 2012), it was some years before occasional research teams began trying to detect marine mammals through eDNA analysis (Székely et al. 2022). In the laboratory, we decided to study the effectiveness and reliability of the approach in various conditions in the field, using reproducible experimental protocols. It seemed obvious to conduct these tests in collaboration with the *Office français de la biodiversité* (OFB – French Biodiversity Agency) in marine protected areas, where it was problematic to monitor relevant marine mammals. We also decided to collaborate with a start-up, Spygen, specializing in the study of eDNA. This collaboration offered multiple benefits: i) Spygen had developed filtration-based seawater sampling protocols, derived from tried-and-tested protocols used in freshwater, and those protocols were ready to be tested in studying marine mammals; ii) Spygen's premises had laboratories which were optimized to minimize false negatives and false positives; and iii) the team used a bio-informatic pipeline which, from the hundreds of thousands of sequences produced for each sample, could perform an initial taxonomic assignment.

Our part of the bargain would be to define the study sites, carry out the sampling, compile a reference database, and perform the final taxonomic assignments. The construction of the reference database and its application to taxonomic assignments was the stage-2 Master's subject worked on by Justine Girardet, who is now a PhD student at the University of Toulon (Girardet 2021).

5.5. First campaign in the Iroise Sea, 2019–2020

The natural marine park of Iroise (PNMI), created by the OFB, was the first natural marine park ever created in France. It is 3500km^2, located on the far west of Finistère, and encapsulates the islands of Ushant, Molène and Sein. Bottlenose dolphins (*Tursiops truncatus*) are present there in great numbers, as are gray seals (*Halichoerus grypus*) and common dolphins (*Delphinus delphis*). Harbor porpoises (*Phocoena phocoena*), which are more difficult to observe, pass through the park, as do Risso's dolphins (*Grampus griseus*), and even some representatives of the mysticeti. There is a pressing need for knowledge about the marine megafauna. Alongside Cécile Gicquel, the coordinator of "natural heritage" projects at the PNMI, and the whole field team, we devised a project hinging on the detection of marine mammals by means of eDNA within the PNMI, also including a test to detect elasmobranchii and bony fish.

Two "positive control" sites (the Chaussée de Sein and Roche des sables in western Bozmen), at which samples were taken in the presence of marine mammals, served to validate the experimental approach (see Figure 5.1). Four test sites were then defined (Pierres Noires, Douarnenez Bay, the coast of Pointe Saint-Mathieu and the Fosse d'Ouessant), along with two sites in Brest Roadstead (Île Ronde and Aulne; where the notable aim was to test the detection of elasmobranchii). In total, sixteen samples were taken, analyzed, and the final sequences taxonomically assigned.

DNA sequences corresponding to mammals were detected in all samples. Six samples showed the recurrent presence of harbor porpoises in the PNMI for the whole duration of the summer, though this animal's recurrent presence was not identified by visual observation. A sequence corresponding, in all likelihood[2], to the bottlenose dolphin was identified in all samples, strongly suggesting that the species used all of these sites. Two DNA sequences, each detected at only one site, cannot be assigned to a precise species: one corresponds to a member of the delphininae subfamily,

2 Taxonomic assignments may be complex in certain cases, such as with the subfamily delphininae. It is particularly important to be able to distinguish similar species such as bottlenose dolphins, common dolphins and blue and white dolphins within the PNMI, but this is a distinction which is often complex to make. This difficulty is inherent in the Delphininae group, whose taxonomy is complex, and in which the species have originated recently and expanded rapidly, which leads to difficulties in species discrimination (Alfonsi et al. 2013; Perrin et al. 2013).

and could therefore come from a, for example, common dolphin or a striped dolphin; the other is similar to a group of three species, the white-beaked dolphin (*Lagenorhynchus albirostris*) being the most likely of the three in Iroise. One final sequence is particularly interesting, though it did not produce a taxonomic assignment: it was NUMT (nuclear mitochondrial DNA), from the integration, into the nuclear genome, of fragments of mitochondrial DNA, whose sequences have evolved separately from the mitochondrial genome since their integration. It is 100% identical to an uncharacterized sequence from the nuclear genome of the Pacific white-sided dolphin (*Lagenorhynchus obliquidens*), but very different from the known mitochondrial DNA sequences from that species.

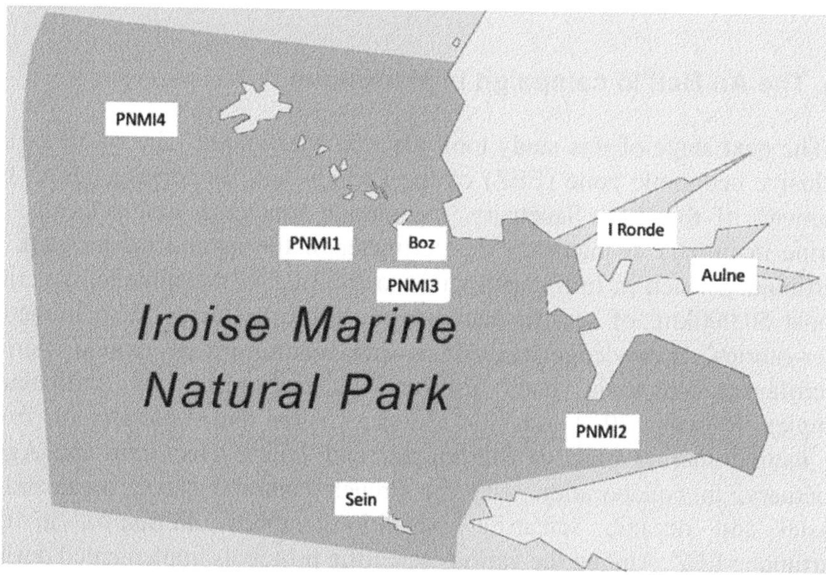

Figure 5.1. *Distribution of the sampling sites in the Iroise Sea and in Brest Roadstead. Sein: chaussée de Sein; Boz: western Bozmen, Roche des sables; PNMI1: Pierres Noires, PNMI2: Douarnenez Bay; PNMI3: off the coast of Pointe Saint-Mathieu; PNMI4: Fosse d'Ouessant; île Ronde (Brest Roadstead), 7: Aulne. For a color version of this figure, see www.iste.co.uk/nicolas/biodiversity.zip*

Two samples were taken in Brest Roadstead in order to test the effectiveness of the method in detecting another taxonomic group: cartilaginous fish. The exact choice of sampling sites (Île Ronde and Aulne – see Figure 5.1) was made thanks to the expertise of Eric Stéphan, of APECS, concerning this taxonomic group. Seven DNA sequences corresponding to elasmobranchii were detected, yielding definite identification for the small-spotted catshark (*Scyliorhinus canicula*), marbled electric ray (*Torpedo marmorata*), and near-certain identification for the undulate ray (*Raja undulata*) and the starry smooth-hound (*Mustelus asterias*).

A coastal sample was taken at Calgrac'h in northern Ushant, serving as a test on the use of another pair of DNA amplifying primers, with broader specificity, targeting all vertebrates. This sample was taken in proximity to a group of harbor porpoises. In it, we detected sequences corresponding to two mammalian taxa (the harbor porpoise, and a species of delphininae, probably *Delphinus delphis* or *Tursiops truncatus*), and 13 taxa of bony fish (10 determined to a species level and three to the level of genus or family).

5.6. The An Bad'lo campaign in Martinique

The next stage of this study took place in Martinique. The whole of the exclusive economic zone (EEZ) of the French Antilles corresponds to the perimeter of the Agoa Sanctuary – a marine protected area dedicated to marine mammals. It intersects the perimeter of the natural marine park of Martinique, which in turn corresponds to the EEZ of Martinique, covering almost 50,000km^2 of sea. In these marine protected areas, with immense "mer-ritories", knowledge acquisition and monitoring of mobile marine macrofauna, mammals, bony fish and elasmobranchii are particularly complex. Between 2019 and 2020, Jessica Crillon and Aude Brador, from the natural marine park of Martinique, and Laurie Hec from the Agoa Sanctuary, in collaboration with the French National Navy, organized a coastal and oceanic scientific expedition dubbed *An Bad'lo*, in the Martinique EEZ. Among the various scientific protocols implemented during this expedition was a biodiversity analysis looking at marine macrofauna (mammals, bony fish and elasmobranchii), as reflected by the study of the eDNA contained in the surface seawater. Samples were taken at nine sites (two coastal sites and seven oceanic sites) from the National Navy's vessel, *Dumont-d'Urville*, in December 2020, with the participation of Étienne Bezault from the University of the French Antilles, and at four coastal sites in Martinique in December 2019 and June 2021 with the help of Benjamin de Montgolfier's team, from the research outfit Aquasearch. The protocol used, from sampling to taxonomic assignment of the sequences obtained, was similar to that used in Iroise.

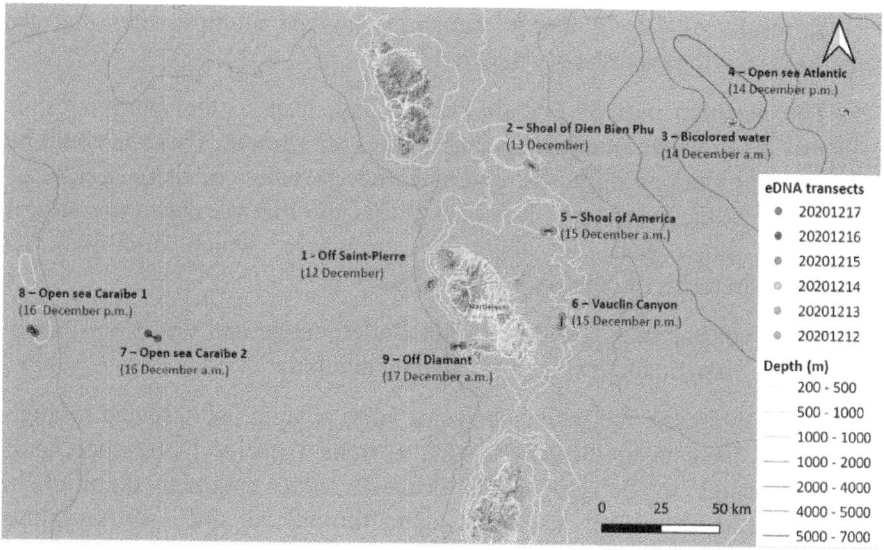

Figure 5.2. *Location, name, sampling date and depth at the nine sites studied during the An Bad'lo campaign (source: Aurore Feunteun, Aquasearch). For a color version of this figure, see www.iste.co.uk/nicolas/biodiversity.zip*

Three species of cetaceans were detected: the abundant presence of pantropical spotted dolphins (*Stenella attenuata*) and Fraser's dolphins (*Lagenodelphis hosei*) was detected at all coastal sites on the Caribbean coast of Martinique, and a taxon belonging to the Delphininae subfamily, but not identifiable to the exact species (this subfamily includes, for example, the bottlenose dolphin *Tursiops truncatus*) was detected at two oceanic sites, notably in particular abundance at the open sea *Atlantic* site. Significant levels of presence of spotted eagle rays (*Aetobatus narinari*) and pelagic stingrays (*Pteroplatytrygon violacea*) were detected in the Fort-de-France Bay and nearby, which confirms recent observations made from the air and on dives (Safi et al. 2021). No other samples showed the presence of elasmobranchii, rays or sharks during the *An Bad'lo* campaign. With regard to bony fish, 193 taxa were identified, of which more than a third were able to be assigned to a particular species (Riou 2022). This corresponds to over half the species expected in Martinique (Bouchon et al. 2002). Coastal and reef-dwelling fish, along with certain pelagic species, represent the vast majority of the taxa detected in the coastal samples. In the open ocean, mainly pelagic fish and deep-water fish displaying nycthemeral migrations were detected.

The results of the *An Bad'lo* campaign showed multiple sites that are remarkable for the biodiversity they hold:

– The northern part of Fort-de-France Bay, near to the mangroves, is evidently a site where multiple species of rays are present. This site could be used for reproduction purposes, and/or harbor juveniles of these species as they grow. At the same time, it is also at this site that we detect the largest number of bony fish, not identified elsewhere, which highlights its importance in terms of conservation efforts.

– Other coastal sites, such as Carbet Bellefontaine and Off Saint-Pierre, also display remarkable levels of taxonomic diversity.

– In the open sea, the main species of large pelagic fish targeted by high sea fishing have been identified, such as tuna (species to be specified), marlin (species to be specified), skipjack tuna, common dolphinfish, barracuda and swordfish. However, again through all open sea sampling campaigns, no elasmobranchii were detected, and only one species of Delphininae was identified, at two of these sites. This result, which is especially important in terms of conservation, strongly suggests that, whilst there is enormous taxonomic diversity of megafauna in the French Antilles, the abundance of cetaceans and elasmobranchii are certainly significantly reduced in the region.

5.7. Detection of marine mammals and other mobile marine fauna through the study of eDNA: from naturalist inventories to the support of public policies

The impacts of environmental change, both at local and global level, are increasingly becoming irreversible. However, a significant portion of marine biodiversity has yet to be explored. For bony and cartilaginous fish, eDNA is now among the recognized monitoring tools (Boussarie et al. 2018; Fernandez et al. 2021; Polanco et al. 2021; Sanchez et al. 2022; Xiong et al. 2022), but the same is not yet true for mammals. The objective of the two research campaigns described in this chapter was to demonstrate the suitability of approaches based on eDNA analysis in marine environments for monitoring the presence of mammals and other representatives of mobile marine fauna.

The effectiveness of the experimental protocol used has been demonstrated – in particular, in Iroise: all of the samples contained detectable eDNA corresponding to mammals, and the species detected vary from one sample to another, thus demonstrating that mammalian eDNA does not tend to become homogenized in the environment. The rate of detection of harbor porpoises with eDNA, which was unexpected in view of the scant visual observations of that species during the same period, underscores the benefit of using this approach when dealing with species that are difficult to study. In the Antilles, for the most part, cetaceans were detected mainly in samples from coastal waters. These results, in combination with those obtained in Iroise, highlight the inabundance of cetaceans in the ocean off Martinique. They are also in keeping with the visual observation of cetaceans by the *An Bad'lo* campaign which, during the seven days at sea, yielded only very rare observations (Benjamin de Montgolfier, Aquasearch, personal communication).

One point on which improvements remain to be made is the processes of taxonomic assignment within the subfamily Delphininae. It is particularly important to be able to distinguish between the close species which are, for example, bottlenose dolphins, common dolphins and striped dolphins. In particular, this will involve defining new DNA sequences to be studied (new markers), or working on multiple markers in parallel.

At present, our work is continuing in this direction. We are also participating in efforts to draw up eDNA standards and data banking. Notably, the aim is to define usage protocols for fauna inventories compiled through eDNA in a marine environment in order to serve the need for monitoring and assessment stemming from public policies, and from the increasing use of the marine environment for renewable energy generation.

5.8. References

Alfonsi, E., Méheust, E., Fuchs, S., Carpentier, F.-G., Quillivic, Y., Viricel, A., Hassani, S., Jung, J.-L. (2013). The use of DNA barcoding to monitor the marine mammal biodiversity along the French Atlantic coast. *ZooKeys*, 365, 5–24. doi: 10.3897/zookeys.365.5873.app2.

Bouchon, C., Bouchon-Navaro, Y., Louis, M. (2002). Les écosystemes marins des Antilles. In *La pêche aux Antilles*, Blanchet, G., Gobert, B., Guererat, J.A. (eds). IRD Editions, La Réunion, 21–44.

Boussarie, G., Bakker, J., Wangensteen, O.S., Mariani, S., Bonnin, L., Juhel, J.-B., Kiszka, J.J., Kulbicki, M., Manel, S., Robbins, W.D. et al. (2018). Environmental DNA illuminates the dark diversity of sharks. *Science Advances*, 4(5), eaap9661. doi: 10.1126/sciadv.aap9661.

Cazau, D., Adam, O., Aubin, T., Laitman, J.T., Reidenberg, J.S. (2016). A study of vocal nonlinearities in humpback whale songs: From production mechanisms to acoustic analysis. *Scientific Reports*, 1–12. doi: 10.1038/srep31660.

Coché, L., Arnaud, E., Bouveret, L., David, R., Foulquier, E., Gandilhon, N., Jeannesson, E., Le Bras, Y., Lerigoleur, E., Lopez, P.J. et al. (2021). Kakila database: Towards a FAIR community approved database of cetacean presence in the waters of the Guadeloupe Archipelago, based on citizen science. *Biodiversity Data Journal*, 9, e69022. doi: 10.3897/bdj.9.e69022.

Fernández, A.P., Marques, V., Fopp, F., Juhel, J., Borrero-Pérez, G.H., Cheutin, M., Dejean, T., Corredor, J.D.G., Acosta-Chaparro, A., Hocdé, R. et al. (2021). Comparing environmental DNA metabarcoding and underwater visual census to monitor tropical reef fishes. *Environmental DNA*, 3(1), 142156. doi: 10.1002/edn3.140.

Foote, A.D., Thomsen, P.F., Sveegaard, S., Wahlberg, M., Kielgast, J., Kyhn, L.A., Salling, A.B., Galatius, A., Orlando, L., Gilbert, M.T.P. (2012). Investigating the potential use of environmental DNA (eDNA) for genetic monitoring of marine mammals. *PLOS ONE*, 7(8), e41781. doi: 10.1371/journal.pone.0041781.t002.

Foote, A.D., Vijay, N., Avila-Arcos, M., Durban, J.W., Fumagalli, M., Gibbs, R.A., Baird, R.W., Hanson, M.B., Korneliussen, T.S., Martin, M.D. et al. (2016). Genome-culture coevolution promotes rapid divergence of killer whale ecotypes. *Nature Communications*, 7, 1–12. doi: 10.1038/ncomms11693.

Girardet, J. (2021). Détection et identification des mammifères marins par analyse d'ADN environnemental. Master's Thesis, Université de Brest, Brest.

Girardet, J., Sarano, F., Richard, G., Tixier, P., Guinet, C., Alexander, A., Sarano, V., Vitry, H., Preud'homme, A., Heuzey, R. et al. (2022). Long distance runners in the marine realm: New insights into genetic diversity, kin relationships and social fidelity of indian ocean male sperm whales. *Frontiers in Marine Science*, 9, 815684. doi: 10.3389/fmars.2022.815684.

Hebert, P.D.N., Cywinska, A., Ball, S.L., deWaard, J.R. (2003). Biological identifications through DNA barcodes. *Proceedings of the Royal Society B: Biological Sciences*, 270(1512), 313–321. doi: 10.1098/rspb.2002.2218.

Jung, J.-L. (ed.) (2017). Approches moléculaires pour l'étude de la biodiversité des mammifères marins. HDR Thesis, Université de Brest, Brest.

Jung, J.-L. and Madon, B. (2021). Protection des mammifères marins face aux activités humaines et nouvelles connaissances issues des études de l'DNA. In *Le transport maritime et la protection de la biodiversité : actes de colloque de Brest, 12 et 13 décembre 2019*, Queffelec, N.B. and Queffelec, B. (eds). Edition Pedone, Paris, 115–146.

Jung, J.-L., Girardet, J., Bonneron, A., Dixneuf, S., Gicquel, C. (2022). Prélèvements d'DNA non invasifs par écouvillonnage de grands dauphins en milieu naturel dans le Parc naturel marin d'Iroise : un test réussi ! *Scientific Reports of Port-Cros National Park*, 36, 213–217.

Morin, P.A., Baker, C.S., Brewer, R.S., Burdin, A.M., Dalebout, M.L., Dines, J.P., Fedutin, I., Filatova, O., Hoyt, E., Jung, J.-L. et al. (2016). Genetic structure of the beaked whale genus *Berardius* in the North Pacific, with genetic evidence for a new species. *Marine Mammal Science*, 1–16. doi: 10.1111/mms.12345.

Perrin, W.F., Rosel, P.E., Cipriano, F. (2013). How to contend with paraphyly in the taxonomy of the delphinine cetaceans? *Marine Mammal Science*. doi: 10.1111/mms.12051.

Polanco, A., Richards, E., Flück, B., Valentini, A., Altermatt, F., Brosse, S., Walser, J., Eme, D., Marques, V., Manel, S. et al. (2021). Comparing the performance of 12S mitochondrial primers for fish environmental DNA across ecosystems. *Environmental DNA*, 3(6), 1113–1127. doi: 10.1002/edn3.232.

Richard, K.R., McKarrey, S., Wright, J.M. (1994). DNA sequence from the SRY gene of sperm whale (*Physeter macrocephalus*) for use in molecular sexing. *Canadian Journal of Zoology*, 72(5), 873–877.

Richard, G., Titova, O.V., Fedutin, I.D., Steel, D., Meschersky, I.G., Hautin, M., Burdin, A.M., Hoyt, E., Filatova, O.A., Jung, J.-L. (2018). Cultural transmission of fine-scale fidelity to feeding sites may shape humpback whale genetic diversity in Russian Pacific waters. *Journal of Heredity*, 25, 2754–11. doi: 10.1093/jhered/esy033Richard.

Riou, M. (2022). Analyse de la biodiversité de la macrofaune marine en Martinique : attributions taxonomiques à partir de données issues de prélèvements d'DNA environnemental. Master's Thesis, Université de Brest, Brest.

Rosel, P.E., Wilcox, L.A., Yamada, T.K., Mullin, K.D. (2021). A new species of baleen whale (*Balaenoptera*) from the Gulf of Mexico, with a review of its geographic distribution. *Marine Mammal Science*, 37(2), 577–610. doi: 10.1111/mms.12776.

Safi, M., Leblanc, H., Cotonea, G., Duporge, G., Feunteun, A., Jung, J.-L., Montgolfier, B. (2021). Identification et inventaire des populations de raies dans la baie de Fort-de-France (Martinique). Report, Parc Naturel Marin de Martinique/Sanctuaire AGOA/Office Français de la Biodiversité.

Sanchez, L., Boulanger, E., Arnal, V., Boissery, P., Dalongeville, A., Dejean, T., Deter, J., Guellati, N., Holon, F., Juhel, J.-B. et al. (2022). Ecological indicators based on quantitative eDNA metabarcoding: The case of marine reserves. *Ecological Indicators*, 140, 108966. doi: 10.1016/j.ecolind.2022.108966.

Sarano, V., Sarano, F., Girardet, J., Prud'Homme, A., Vitry, H., Heuzey, R., Sarano, M., Delfourt, F., Glotin, H., Adam, O. et al. (2022). Underwater photo-identification of sperm whales (*Physeter macrocephalus*) off Mauritius. *Marine Biology Research*, 18(1/2), 131–146. doi: 10.1080/17451000.2022.2040737.

Székely, D., Cammen, K.M., Olsen, M.T. (2022). Needles in an ocean haystack: Using environmental DNA to study marine mammals in the North Atlantic. *NAMMCO Scientific Publications*, 12. doi: 10.7557/3.6482.

Taberlet, P., Bonin, A., Zinger, L., Coissac, E. (2018). *Environmental DNA: For Biodiversity Research and Monitoring*. Oxford University Press, Oxford.

Valentini, A., Pompanon, F., Taberlet, P. (2009). DNA barcoding for ecologists. *Trends in Ecology & Evolution*, 24(2), 110–117. doi: 10.1016/j.tree.2008.09.011.

Whitehead, H. (2017). Gene–culture coevolution in whales and dolphins. *Proceedings of the National Academy of Sciences of the United States of America*, 114(30), 7814–7821. doi: 10.1073/pnas.1620736114.

Xiong, F., Shu, L., Zeng, H., Gan, X., He, S., Peng, Z. (2022). Methodology for fish biodiversity monitoring with environmental DNA metabarcoding: The primers, databases and bioinformatic pipelines. *Water Biology and Security*, 1(1), 100007. doi: 10.1016/j.watbs.2022.100007.

6

DNA Barcoding for Identifying Species and Monitoring French Biodiversity

6.1. Introduction

In the context of the global biodiversity crisis, and growing recognition of its impact on human societies (ecosystem services, harmful agents, invasive species, zoonosis transmission, etc.), the stakes linked to species identification and to improving biodiversity-monitoring capabilities are universally recognized (IPBES 2019; Lévêque and Witté 2019). Developing these capabilities for a better understanding of the impact of environmental changes on species is a major objective for both national and international programs. This involves the implementation of innovative and ambitious approaches to address the challenges imposed by the spatial and temporal scales that need to be taken into account, but also to be able to integrate highly diverse and abundant components of biodiversity, which are often not well known and have largely been ignored to date. Such is the case, for example, with insects, which are essential to the functioning of ecosystems (recycling of organic material; a key link in the food chain for many other organisms; vectors of parasites, pathogenic bacteria, viruses; crop

---

Chapter written by Rodolphe ROUGERIE, Lucas SIRE and Antoine LÉVÊQUE.

auxiliaries; etc.). However, their declining populations and diversity have only begun to be recognized and taken into account in recent years (Wagner et al. 2021).

The development of standardized genetic approaches to species identification (DNA barcoding) since the early 2000s (Hebert et al. 2003) has provided an alternative method to traditional, mainly morphological, techniques. The benefits of these molecular methods are significant because, firstly, identification can be performed whatever the DNA source used, and therefore, whatever developmental stage the organism is at (juvenile, egg, etc.) or the type of sample available (insect leg, hair, blood, etc.) and, secondly, the identification does not require expert taxonomists, in a context where these skills are in short supply (Engel et al. 2021).

The constant development of molecular-biology and DNA-sequencing technologies has made these methods more widely accessible over the course of recent decades. In particular, the advent of new high-throughput sequencing (HTS) technologies has considerably reduced costs, and facilitated the development of so-called "metagenomic" methods, whose aim is to sequence DNA not from a single individual, but from all individuals present in a sample (Porter and Hajibabaei 2018). Thus, with an approach known as "metabarcoding", it is possible to characterize all of the species present in a complex sample in order to document, say, the species within a community, or all of the organisms represented in the environmental DNA (eDNA) extracted from a sample of water, soil or even air (Taberlet et al. 2012; Clare et al. 2022). However, the power of these approaches is still conditional on the existence, exhaustiveness and reliability of reference libraries of DNA sequences, previously compiled by expert taxonomists (Ekrem et al. 2007).

Considerable progress has been made across a wide range of taxa and regions of the globe (e.g. for fish: Delrieu-Trottin et al. 2019; birds: Kerr et al. 2007; insects: deWaard et al. 2019; Dinca et al. 2021; Hendrich et al. 2014; Lopez-Vaamonde et al. 2021; and plants: Setsuko et al. 2023) and the DNA barcode database (BOLD, *The Barcode of Life Datasystems* (Ratnasingham and Hebert 2007)) now includes over 13 million DNA sequences representing around 350,000 identified species.

6.2. DNA barcoding for species identification

6.2.1. *A simple and standardized tool for molecular identification*

Primarily developed in the field of microbiology around the late 20th century, molecular identification of species experienced a major boost following the proposal, in 2003, by Canadian researcher Paul Hebert and his colleagues, to generalize a standardized approach to eukaryotic species identification called DNA barcoding (Hebert et al. 2003). The technique is based on the comparison of short DNA sequences of a genetic marker, called a *DNA barcode*, in reference to the barcodes that are used worldwide to identify products. Proposing to use a standard DNA fragment whose sequences vary considerably between distinct species, but only very slightly between individuals of the same species, the underlying idea is that of a world where, by means of DNA barcoding, all animal, plant and fungal species would be easily identifiable, and the inventory of biodiversity would thus be simplified and accelerated.

The DNA marker initially proposed by Hebert et al. (2003) – a fragment of around 650 base pairs of the mitochondrial COI (Cytochrome c Oxidase, subunit I) gene – was empirically tested in numerous taxonomic groups, and subsequently adopted as a standard DNA barcode in animals. In addition, international communities of researchers agreed on the selection of a multi-marker approach in plants (a combination of two markers: MatK and rbcl, supplemented, depending on the target taxon, by trnH-psbA or ITS (CBOL Plant Working Group 2009; Cheng et al. 2016), and the ITS marker in fungi (Schoch et al. 2012)). Species identification using DNA barcoding, however, has received considerable conceptual criticisms (for example, Moritz and Cicero 2004; Rubinoff and Holland 2005), and was the subject of some controversy linked to the abstraction of a costly molecular taxonomy approach compared to "traditional" morphology-based taxonomic approaches, suffering from a dire lack of human and financial resources (Will et al. 2005; Cameron et al. 2006). However, many taxonomists contributed to the successful development of DNA barcoding and to a much more mature understanding of its potential and its limitations by its use in an integrative context (Goldstein and Desalle 2011); that is, as a complement to classical approaches of comparative morphology. The success of this so-called integrative taxonomy (Dayrat 2005; Padial et al. 2010), combined with the simplicity and increasing accessibility of DNA sequencing thanks to technological developments in molecular biology, led to the adoption, now widespread throughout the world, of DNA barcodes as a relevant tool for species identification and discrimination.

6.2.2. *From the specimen to the community: identification by DNA barcoding and metabarcoding*

In its simplest initial design, proposed in the early 2000s when the dominant DNA-sequencing method was Sanger sequencing (Sanger et al. 1977), a DNA barcode is a simple nucleotide sequence obtained for an individual, representing the standard genetic marker relevant to its identification (for instance, COI for an animal, or MatK+rbcl for a plant (see section 6.2.1)). This sequence is produced either to identify an unknown specimen by comparing it to existing libraries of reference sequences, or for the very construction of these libraries by associating a sampled specimen with a complete and reliable identification by an expert taxonomist (though it may be revised later if necessary). This comparison of sequences for identification is generally carried out using tools implemented directly in the main genetic sequence database, such as the BOLD "identification engine" or GenBank's BLAST tool (Benson et al. 2013[1]). The results come in the form of percentages of similarity between the unknown sequence and the closest sequences in the reference libraries. Thus, the objective is to lead the user to propose an identification hypothesis, the reliability of which is being conditioned on that of the libraries by way of, for example, the metadata linked to the results, which could refer to a scientific publication and/or provide information about the origin of the associated taxonomic knowledge.

The development of HTS, linked with the major human genome sequencing projects, has revolutionized access to genetic data in terms of volume of data produced, speed of implementation and reduction of the associated costs. Thus, it has become possible, at a very high throughput and

1 See: www.blast.ncbi.nlm.nih.gov/.

for a fraction of the cost of Sanger sequencing, to sequence the DNA barcodes of specimens for their identification (Meier et al. 2015; Shokralla et al. 2015; Srivathsan et al. 2021) or for constructing reference libraries (Cruaud et al. 2017; Hebert et al. 2018). In addition, HTS has opened the door to metagenomic approaches, wherein the sequencing target is no longer the DNA of an individual's genome, but that of the combination of genomes or genome fragments of multiple organisms in a complex sample. Using this approach, it is possible, for example, to simultaneously sequence the DNA barcodes of all individual insects contained in the same trap sample (Morinière et al. 2016; Liu et al. 2023); a method called "metabarcoding".

Thus, a list of the species present in a community can be produced thanks to bioinformatic tools dedicated to processing the millions of sequences produced and comparing them against references in order to identify them (Creedy et al. 2022; Hakimzadeh et al. 2023).

Environmental DNA (eDNA) sequencing is a variant of this approach, where the substrate is the sample of the medium itself (e.g. soil, water or air, often sampled by filtration) instead of a complex set of individuals physically present in the sample being processed. The DNA traces left by the organisms can thus be detected and identified by HTS (Apothéloz-Perret-Gentil et al. 2021; Kirse et al. 2021; West et al. 2021; Carvalho et al. 2022; Clare et al. 2022).

Irrespective of the approach used to produce the genetic data necessary for species identification, barcoding or metabarcoding, they ultimately depend on the existence, accessibility and reliability of DNA reference libraries. The construction of these libraries is a "work in progress", which may draw on various strategies targeting taxa (Dinca et al. 2021; Lopez-Vaamonde et al. 2021), geographic zones (deWaard et al. 2019; Setsuko et al. 2023) or habitats (Weigand et al. 2019), drawing upon the taxonomic expertise present in the collections of museums (Santos et al. 2023), and ideally applying rigorous standards in the compilation and management of the metadata associated with the specimens (Rimet et al. 2021).

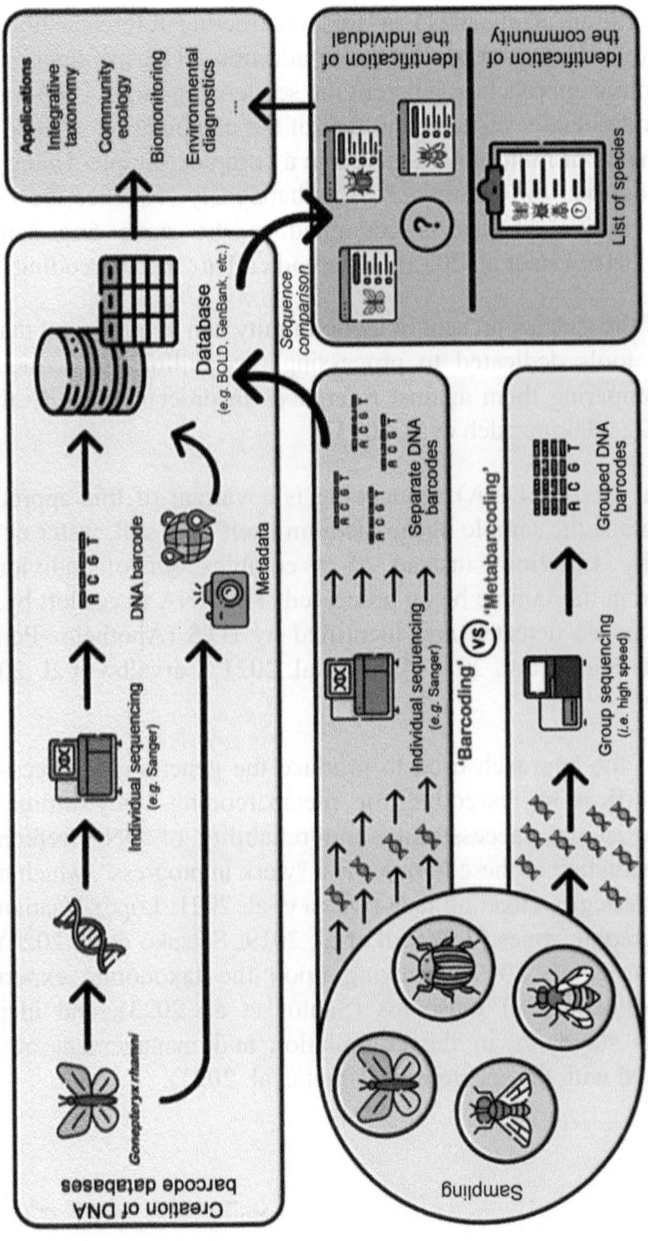

Figure 6.1. *Diagrammatic representation of the process of molecular identification of species by barcoding or metabarcoding (source of images taken by Freepik: www.flaticon.com). For a color version of this figure, see www.iste.co.uk/nicolas/biodiversity.zip*

6.2.3. *DNA barcodes of the living world: a collective and international effort*

The interest sparked by the Hebert et al. (2003) proposal soon gained a structure within an international community of biologists, pursuing this idea of large-scale standardization of molecular identification of species. The work of this community and the organization of two-yearly international conferences (the first being held in London in 2005) facilitated the construction of a solid empirical foundation, built on vast quantities of data for numerous taxa, in various regions of the world, and for applications in different disciplines. Boosted by major funding in Canada, the first phase of the iBOL (international Barcode of Life, see: www.ibol.org) program, called *barcode 500K*, was launched in 2010 under the leadership of the University of Guelph (Ontario). It represented a total investment of nearly $150 million, with 25 participating nations[2], and ended five years later, having achieved the shared objective of constructing reference libraries of DNA barcodes for 500,000 species of eukaryotic multicellular organisms. In Europe, major national projects, such as the *German Barcode of Life* (GBOL)[3], the *Finnish Barcode of Life* (FinBOL)[4] or France's "Bibliothèque du Vivant" project (Le Gall et al. 2017), contributed to the success of the program. This first phase also saw the development of dedicated infrastructures in Canada, with, firstly, a molecular biology laboratory specially devoted to the high-throughput generation of DNA barcodes (Canadian Centre for DNA barcoding[5]) and, secondly, a common Web-based platform for the storage, handling and analysis of barcoding data: BOLD (see Figure 6.2) (Ratnasingham and Hebert 2007), developed at the Centre for Biodiversity Genomics[6] in Guelph.

Building on this collective accomplishment and solid infrastructures, the international DNA barcoding community has committed to a second

2 See: www.ibol.org/phase1/about-us/partner-nations/.

3 See: www.bolgermany.de.

4 See: www.finbol.org.

5 See: www.ccdb.ca.

6 See: www. biodiversitygenomics.net.

phase for the period 2019–2028, called BIOSCAN[7]. The objective, pursued by over 30 participating nations, is threefold:

– to continue the documentation and discovery of biodiversity through the development of DNA reference libraries for two million species;

– to reveal the interactions between species by analyzing their symbiomes;

– to implement, in at least 2000 sites across all continents, systems dedicated to monitoring biodiversity through the molecular analysis of communities.

Once again, major funding obtained in different countries has enabled the launch of this program, the total cost of which is estimated to be $180 million. This is the case in Europe, where national initiatives are being developed, and where the community has organized itself within the BIOSCAN Europe[8] consortium, recently dynamized by the support of the European Union for the Biodiversity Genomics Europe consortium.[9]

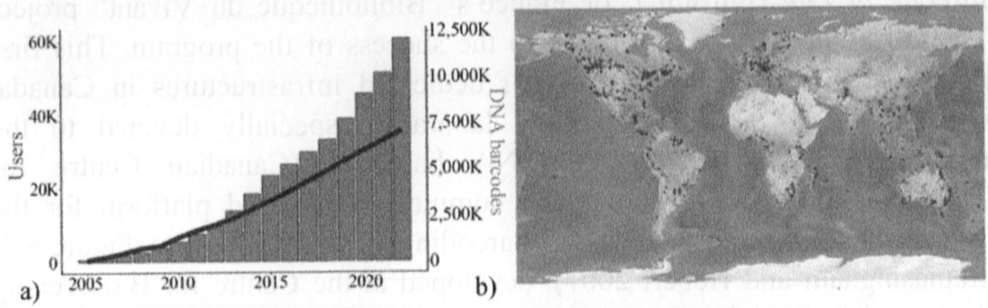

Figure 6.2. *Development of BOLD since its creation in 2005: a) evolution of the number of registered users (blue line) and specimens sampled (brown bars); b) geographical distribution of records in BOLD (as at May 5, 2023[10]). For a color version of this figure, see www.iste.co.uk/nicolas/biodiversity.zip*

At the time of writing (in July 2023), BOLD holds the DNA barcodes of almost 13.5 million specimens, representing 250,000 identified species of animals, 72,000 plant species and 25,000 fungi, but also a large number of

7 See: www.ibol.org/site/programs/bioscan/.

8 See: www.bioscaneurope.org.

9 See: www.biodiversitygenomics.eu.

10 Data disclosed by boldsystems.org.

unidentified species. In BOLD, an automatic registry discriminating between clusters of DNA barcodes (called Barcode Index Numbers or BINs (Ratnasingham and Hebert 2013)) as molecular proxies for the species, currently lists almost 950,000 BINs in total.

6.3. DNA barcode libraries for biodiversity in Metropolitan France and overseas territories

France, as a geographic entity comprising Metropolitan France and the French overseas territories (known as DROM-COM), contains a wide variety of territories and ecosystems, with rich and sometimes unique biodiversity (for example, New Caledonia, Réunion Island and Corsica harbor many endemic species). According to the National Inventory of Natural Heritage (INPN[11]), nearly 200,000 species (fauna, flora, fungi and micro-organisms, across all environments) live in Metropolitan France and its overseas territories, including around 150,000 terrestrial species. However, our knowledge of French biodiversity is incomplete, and largely skewed at the expense of certain taxa such as arthropods (especially small ones) and territories where biodiversity is the greatest (for example, French Guiana). To catalog this biodiversity, the INPN has developed a taxonomic repository, TAXREF (v16.0, see Gargominy et al. 2022), which provides an up-to-date catalog of known French species within a rigorous nomenclatural framework. By cross-referencing the taxonomic and geographical information contained in this repository with those from the DNA barcode reference libraries in BOLD, it is possible to assess the advances in capabilities for molecular identification of French species (see Figure 6.3 and Table 6.1).

A little under half (86,712 (43%)) of the 203,367 eukaryotic species listed in TAXREF are currently represented by at least one DNA barcode in BOLD, and only around a quarter (51,920 (26%)) are represented by at least five barcodes, and therefore by a certain degree of intraspecies genetic variability. This is an important parameter in assessing the relevance of DNA barcoding for species identification. Whilst the vast majority of these DNA barcodes in BOLD are publicly accessible (75,301 (87%)), it is possible

11 See: www.inpn.mnhn.fr/.

to search for them all using the BOLD ID engine[12], in which one or more unknown sequences can be used to query these libraries.

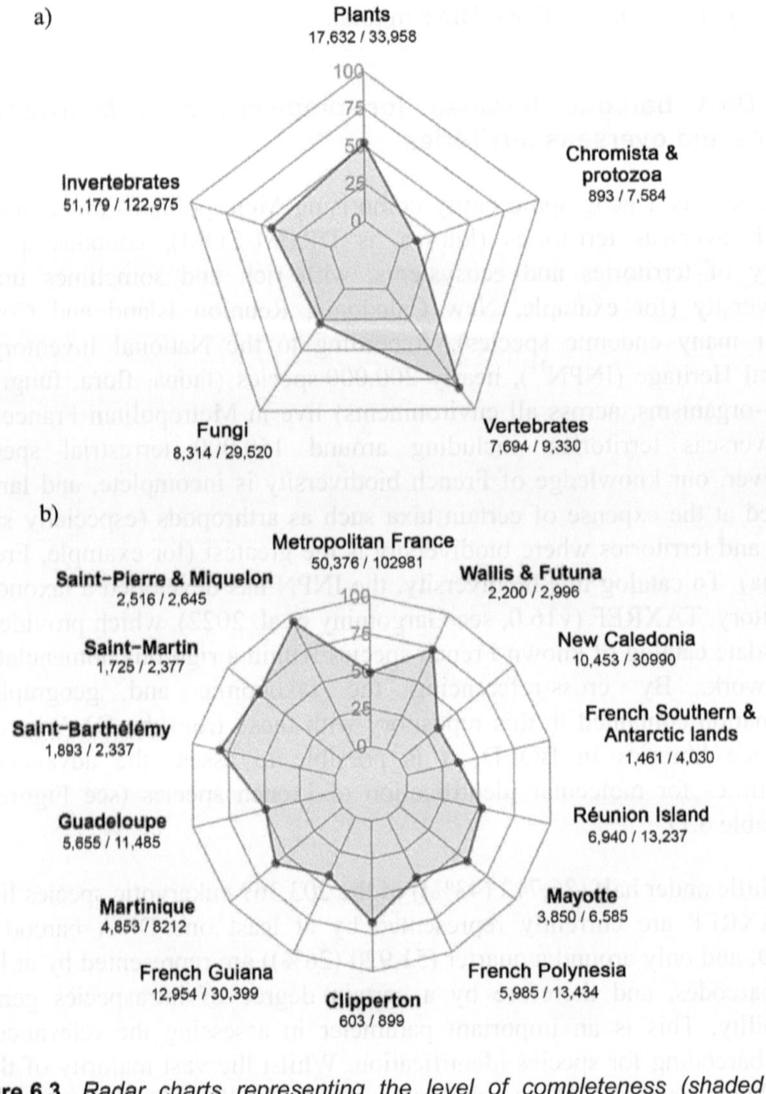

Figure 6.3. *Radar charts representing the level of completeness (shaded area, expressed in %) of DNA barcode libraries in BOLD for French eukaryotic biodiversity: a) by taxonomic group; b) by geographic region. The values quoted (—/—) for each category represent the number of species represented in BOLD and the total number in TAXREF v16.0, respectively*

12 See: http://v4.boldsystems.org/index.php/IDS_OpenIdEngine.

With respect to taxonomy, vertebrates have the best coverage in BOLD, with 82% of species represented by at least one DNA barcode, compared to only 42% for invertebrates. This reflects the general level of maturity of taxonomic knowledge of vertebrate species (Burgin et al. 2018), their popularity with the public and the scientific community (for example, in conservation biology). It is also indicative of the significant disproportion between the diversity of these taxonomic groups and the effort put into studying them (Beck et al. 2012). In invertebrates, these values should also be tempered by the large number of species that have yet to be discovered and described, particularly among insects and in tropical territories (Lees et al. 2014; Stork et al. 2018). It should also be noted that the diversity of these territories is greatly underestimated by the current contents of TAXREF, which records more species for Metropolitan France alone than for all 13 DROM-COMs combined, which assuredly does not reflect the actual distribution of species diversity within the French territories.

From a geographical point of view, the completeness of DNA barcode databases varies from one territory to another. It is generally high in the least diverse DROM-COMs (for example, it is 95% complete for Saint-Pierre and Miquelon, and 81% complete for Saint-Barthélemy). In contrast, it is low for territories where biodiversity is high, little known and sometimes highly endemic (for example, Réunion Island (52%), French Guiana (43%), French Polynesia (45%), and New Caledonia (34%)) (see Figure 6.3). While the overall coverage of French eukaryotic biodiversity (43%) is still insufficient, it is even more so when the geographical origin of the sequenced specimens is taken into account. Indeed, this coverage is no more than 11% if we consider only the species represented by at least one specimen sampled on French territory, because specimens from neighboring countries are largely responsible for the representation of French species in BOLD.

Because TAXREF is incomplete for certain territories, it is difficult to assess the effort needed to achieve exhaustive coverage of French biodiversity. On the other hand, for Metropolitan France, where taxonomic knowledge and the completeness of TAXREF are highly advanced, the current state of coverage of DNA barcode libraries in BOLD (see Table 6.1) is representative, and can be used to gage the effort needed – still quite substantial – to achieve coverage that would at least document a certain amount of intraspecific genetic variability for local populations sampled in France. At present, fewer than 40% of invertebrate species are represented by at least five DNA barcodes, and only 16% by specimens from France. Only slightly less than a quarter of vertebrate species are represented by specimens sampled in Metropolitan France, and this falls to 6% and 2% in the case of plants and fungi, respectively. In total, though, nearly half (50,376 (49%)) of species in Metropolitan France are represented in BOLD, and the majority are represented by DNA barcodes and records whose metadata are publicly available.

	TAXREF	BOLD (1+)	BOLD (5+)	BOLD (1P)	BOLD (FR)
Plants	11,064	6,486 (59%)	4,243 (38%)	6,254 (57%)	694 (6%)
Fungi	25,254	7,504 (30%)	3,223 (13%)	6,342 (25%)	499 (2%)
Chromista, protozoa	4,424	719 (16%)	272 (6%)	673 (15%)	54 (1%)
Vertebrates	1,482	1,391 (94%)	1,257 (85%)	1,359 (92%)	405 (27%)
Invertebrates	60,757	34,276 (56%)	23,780 (39%)	29,384 (48%)	9,983 (16%)
Total	102,981	50,376 (49%)	32,775 (32%)	44,012 (43%)	11,635 (11%)

Table 6.1. *Overview, for the five main groups of eukaryotes considered, of the number of validated species (TAXREF) for Metropolitan France, and the number of species are represented in BOLD*

COMMENT ON TABLE 6.1.– *At least one (1+) or at least five (5+) DNA barcodes; at least one publicly available DNA barcode (1P) or at least one DNA barcode derived from a specimen found in Metropolitan France (FR). The associated percentages refer, in each case, to the total number of species listed in TAXREF*.

6.4. Main challenges for the molecular identification of species in France

6.4.1. *Development of exhaustive and accessible reference libraries*

The state of the art presented in the previous section demonstrated the scale of the efforts still required to meet the major challenge of reaching exhaustiveness of DNA barcode reference libraries, especially with the aim of inclusion of populations originating from French territories, but also with regard to the representativeness of intraspecific variability for the genetic marker used. Several priority action plans can be put forward:

– a taxonomy-centered approach, with priority given to the species still absent from BOLD, or, with a lower degree of priority, those only represented in BOLD by samples originating from non-French territories;

– a geographical focus, to sample massively in the most diverse and least represented regions (e.g. the Mediterranean Basin, DROM-COMs such as French Guiana or New Caledonia, etc.);

– a focus on status, prioritizing the sampling of populations of endemic, protected or heritage species;

– a socio-economic view, focused on species that provide ecosystem services, and are valued in human activities (e.g. pollinators, crop auxiliaries, and parasitoids used in biological control), or, conversely, species considered harmful (e.g. invasive species, crop pests, and insects transmitting zoonoses).

National coordination of the implementation of these multiple strategies has proved a success in various countries (for example, GBOL in Germany[13]), where campaigns to construct DNA-barcode reference libraries have been carried out. In France, where the community of users of DNA barcodes is well developed, but dispersed geographically (Metropolitan France and the DROM-COMs) and by topic (taxonomists, ecologists, conservation biologists, agronomists, etc.), such coordination would make it possible to better integrate and exploit existing activities and data, to optimize future sampling campaigns and reduce redundancy, or provide wider access to infrastructures (for example, natural history collections, or molecular biology lab facilities) and to the funding needed to build these libraries. In order to cope with a growing demand for development of our capabilities for molecular identification of species, particularly in academia, and in reference to the problem of biodiversity monitoring (see section 6.4.2), it is also important that this hoped-for development of libraries of DNA barcodes takes place with the guarantee that the data generated will be accessible (sequences, metadata associated with the specimens, images), in accordance with the FAIR principles (Wilkinson et al. 2016) – that is, the data should be Findable, Accessible, Interoperable and Reusable.

---

13 See: www.bolgermany.de.

6.4.2. *Molecular identification of species: an asset for monitoring French terrestrial biodiversity*

The need for monitoring of biodiversity in Europe emerged politically in 1979, with the introduction of the EU "Birds" directive and then, more broadly for the rest of biodiversity, in 1992, with the "Habitats" directive on flora and fauna. A number of biodiversity monitoring schemes, mostly relating to vertebrates and primarily implemented in Metropolitan France by traditional wildlife observation approaches, have been developed by different stakeholders. However, there is no such monitoring being deployed as part of an overarching, structured and coherent program on a national scale (Touroult et al. 2017a, 2017b). Initially formalized in action 70 of the Interministerial Biodiversity Plan dated July 4, 2018, the implementation of terrestrial biodiversity monitoring in France was reaffirmed in the National Biodiversity Strategy 2020–2030, and a national terrestrial biodiversity monitoring program was launched by the Ministry for Ecology, which entrusted its organization to the French Biodiversity Agency (*Office français de la biodiversité* (OFB)). The challenge is to have a continuous flow of up-to-date, reliable data and information on biodiversity that can be compared across time and space, with the aim of monitoring its state, assessing the results of conservation efforts, and guiding future actions of public authorities, and society in general, to make them as effective as possible.

Whilst in practice, biodiversity monitoring can use a range of methods and protocols (for instance, visual or acoustic tracking), the use of molecular identification of species offers new prospects in terms of geographical and temporal scales, or target taxa. High-throughput identification approaches using metagenomics (environmental DNA, metabarcoding) are particularly promising for monitoring insects, whose serious and worrying population decline has only come to light recently (for example, in Germany (Hallman et al. 2017) and Costa Rica (Janzen and Hallwachs 2021)). Metabarcoding analyses of insects caught in so-called "Malaise traps" (Kaczmarek et al. 2022; Sire et al. 2022; Liu et al. 2023) clearly illustrate the contribution of these metagenomic approaches capable of processing a very large number of specimens and species in taxonomic groups for which taxonomic expertise is largely or even entirely lacking. The international program BIOSCAN (see section 6.2.3) and related national (for example, BIOALFA in Costa Rica[14], BIOSCAN-FR in France) or international (for example, LIFEPLAN[15])

14 See: www.gdfcf.org/bioalfa-bioliteracy-costa-rica.

research projects use this type of device to monitor communities of flying insects (see Figure 6.4). More broadly, these molecular approaches and the constant technological innovations associated with them (Pomerantz et al. 2018; Wägele et al. 2022; Chua et al. 2023) could enable us to increase the number of permanent monitoring sites for taxonomic groups (for example, insects, protista, fungi, etc.) and ecosystems such as the soil (Young and Hebert 2022) or even air (Clare et al. 2022) which were previously impossible to implement, or were only partially implemented. In regions of the world where biodiversity is the richest and the least well known, these molecular tools offer the opportunity to take account of as-of-yet unknown and undescribed taxa, known as dark taxa (Hartop et al. 2022), by using the concept of molecular operational taxonomic unit, "MOTU" (Blaxter and Floyd 2003), or a standardized proxy such as BINs (see section 1.3, (Ratnasingham and Hebert 2013)). In the DROM-COMs, where these dark taxa represent a considerable portion of the communities – for example, in Guiana (Lees et al. 2014; Decaëns et al. 2016) – the use of these molecular approaches is thus a possible response to the growing challenge of biodiversity monitoring in areas subject to strong anthropic pressure or with high sensitivity to global changes and to invasive exotic species.

a) b)

Figure 6.4. *Operational test of a device for monitoring insect populations in France as part of the BIOSCAN-FR project. a) A Malaise trap installed in Guiana and operated by the OFB; b) a map of the pilot sites used by the project where one or multiple test devices were deployed in 2022 and 2023. For a color version of this figure, see www.iste.co.uk/nicolas/biodiversity.zip*

15 See: www.helsinki.fi/en/projects/lifeplan.

6.5. Conclusion and perspectives

The use of DNA barcodes and metagenomic approaches represents an innovative solution for species identification in contexts where traditional approaches are difficult to apply (wide geographic area, high number of samples, taxonomic deficit, etc.). The very nature of the genetic information used makes it easy to store so that it can be reused and that identifications can be revisited in light of new results. This offers significant added value to the use of this type of data, notably in the context of monitoring where sampling is continuous over time (Berry et al. 2020). Molecular species identification is thus particularly appropriate for concerns relating to the "FAIRization" of data linked to the documentation and monitoring of biodiversity (Lannom et al. 2020; Royaux et al. 2022).

Today, the major challenges linked to knowledge of biodiversity and its monitoring are universally recognized. The use of these molecular-level approaches to meet them has become the object of ambitious national and international initiatives (see section 6.2.3), in which France is actively participating and contributing (Bonants et al. 2010; Le Gall et al. 2017) (see Observatoire Vigilife[16]). The current development of a national network (FrBOL: "French Barcode of Life"), similar to those existing in other countries and bringing together national actors, both academics and nonacademics, involved in the use of these genetic tools, will notably facilitate the coordination of the ongoing efforts to compile reference libraries (see section 6.3). This network could streamline and generalize access to infrastructures and technical skills, to training in the use of these approaches, and to financial resources. Finally, in the context of biodiversity shared by all, and of global concerns about its knowledge and conservation of it, it will bridge the gap between initiatives and advances at national level and those made by the international community.

6.6. References

Apotheloz-Perret-Gentil, L., Bouchez, A., Cordier, T., Cordonier, A., Gueguen, J., Rimet, F., Vasselon, V., Pawlowski, J. (2021). Monitoring the ecological status of rivers with diatom eDNA metabarcoding: A comparison of taxonomic markers and analytical approaches for the inference of a molecular diatom index. *Mol. Ecol.*, 30(13), 2959–2968.

16 See: www.vigilife.org/.

Beck, J., Ballesteros-Mejia, L., Buchmann, C.M., Dengler, J., Fritz, S.A., Gruber, B., Hof, C., Jansen, F., Knapp, S., Kreft, H. et al. (2012). What's on the horizon for macroecology? *Ecography*, 35(8), 673–683.

Benson, D.A., Cavanaugh, M., Clark, K., Karsch-Mizrachi, I., Lipman, D.J., Ostell, J., Sayers, E.W. (2013). GenBank. *Nucleic Acids. Res.*, 41, D36–42.

Berry, O., Jarman, S., Bissett, A., Hope, M., Paeper, C., Bessey, C., Schwartz, M.K., Hale, J., Bunce, M. (2020). Making environmental DNA (eDNA) biodiversity records globally accessible. *Environ. DNA*, 3(4), 699–705.

Blaxter, M. and Floyd, R. (2003). Molecular taxonomics for biodiversity surveys: Already a reality. *Trends Ecol. Evol.*, 18(6), 268–269.

Bonants, P., Groenewald, E., Rasplus, J.Y., Maes, M., De Vos, P., Frey, J., Boonham, N. Nicolaisen, M. Bertacini, A. Robert, V. et al. (2010). QBOL: A new EU project focusing on DNA barcoding of Quarantine organisms. *EPPO Bull.*, 40(1), 30–33.

Burgin, C.J., Colella, J.P., Kahn, P.L., Upham, N.S. (2018). How many species of mammals are there? *J. Mammal.*, 99(1), 1–14.

Cameron, S., Rubinoff, D., Will, K. (2006). Who will actually use DNA barcoding and what will it cost? *Syst. Biol.*, 55(5), 844–847.

Carvalho, C.S., De Oliveira, M.E., Rodriguez-Castro, K.G., Saranholi, B.H., Galetti Jr., P.M. (2022). Efficiency of eDNA and iDNA in assessing vertebrate diversity and its abundance. *Mol. Ecol. Resour.*, 22(4), 1262–1273.

CBOL Plant Working Group (2009). A DNA barcode for land plants. *Proc. Natl. Acad. Sci. USA*, 106(31), 12794–12797.

Cheng, T., Xu, C., Lei, L., Li, C., Zhang, Y., Zhou, S. (2016). Barcoding the kingdom Plantae: New PCR primers for ITS regions of plants with improved universality and specificity. *Mol. Ecol. Resour.*, 16(1), 138–149.

Chua, P.Y.S., Bourlat, S.J., Ferguson, C., Korlevic, P., Zhao, L., Ekrem, T., Meier, R., Lawniczak, M.K.N. (2023). Future of DNA-based insect monitoring. *Trends Genet.*, 39(7), 531–544.

Clare, E.L., Economou, C.K., Bennett, F.J., Dyer, C.E., Adams, K., Mcrobie, B., Drinkwater, R., Littlefair, J.E. (2022). Measuring biodiversity from DNA in the air. *Curr. Biol.*, 32(3), 693–700.

Creedy, T.J., Andujar, C., Meramveliotakis, E., Noguerales, V., Overcast, I., Papadopoulou, A., Morlon, H., Vogler, A.P., Emerson, B.C., Arribas, P. (2022). Coming of age for COI metabarcoding of whole organism community DNA: Towards bioinformatic harmonisation. *Mol. Ecol. Resour.*, 22(3), 847–861.

Cruaud, P., Rasplus, J.Y., Rodriguez, L.J., Cruaud, A. (2017). High-throughput sequencing of multiple amplicons for barcoding and integrative taxonomy. *Sci. Rep.*, 7, 41948.

Dayrat, B. (2005). Towards integrative taxonomy. *Biol. J. Linn. Soc.*, 85(3), 407–415.

Decaëns, T., Porco, D., James, S.W., Brown, G.G., Chassany, V., Dubs, F., Dupont, L., Lapied, E., Rougerie, R., Rossi, J.-P., Roy, V. (2016). DNA barcoding reveals diversity patterns of earthworm communities in remote tropical forests of French Guiana. *Soil Biology and Biochemistry*, 92, 171–183.

Delrieu-Trottin, E., Williams, J.T., Pitassy, D., Driskell, A., Hubert, N., Viviani, J., Cribb, T.H., Espiau, B., Galzin, R., Kulbicki, M. et al. (2019). A DNA barcode reference library of French Polynesian shore fishes. *Sci. Data*, 6(1), 114.

deWaard, J.R., Ratnasingham, S., Zakharov, E.V., Borisenko, A.V., Steinke, D., Telfer, A.C., Perez, K.H.J., Sones, J.E., Young, M.R., Levesque-Beaudin, V. et al. (2019). A reference library for Canadian invertebrates with 1.5 million barcodes, voucher specimens, and DNA samples. *Sci. Data*, 6(1), 308.

Dinca, V., Dapporto, L., Somervuo, P., Voda, R., Cuvelier, S., Gascoigne-Pees, M., Huemer, P., Mutanen, M., Hebert, P.D.N., Vila, R. (2021). High resolution DNA barcode library for European butterflies reveals continental patterns of mitochondrial genetic diversity. *Commun. Biol.*, 4(1), 315.

Ekrem, T., Willassen, E., Stur, E. (2007). A comprehensive DNA sequence library is essential for identification with DNA barcodes. *Mol. Phylogenet. Evol.*, 43(2), 530–542.

Engel, M.S., Ceríaco, L.M.P., Daniel, G.M., Dellapé, P.M., Löbl, I., Marinov, M., Reis, R.E., Young, M.T., Dubois, A., Agarwal, I. et al. (2021). The taxonomic impediment: A shortage of taxonomists, not the lack of technical approaches. *Zool. J. Linn. Soc.*, 193(2), 381–387.

Gargominy, O., Tercerie, S., Ramage, T., Dupont, P., Daszkiewicz, P., Poncet, L. (2022). TAXREF, référentiel taxonomique pour la France : méthodologie, mise en œuvre et diffusion. Report, MNHN, Paris.

Goldstein, P.Z. and Desalle, R. (2011). Integrating DNA barcode data and taxonomic practice: Determination, discovery, and description. *Bioessays*, 33(2), 135–147.

Hakimzadeh, A., Asbun, A.A., Albanese, D., Bernard, M., Buchner, D., Callahan, B., Caporaso, J.G., Curd, E., Djemiel, C., Durling, B. et al. (2023). A pile of pipelines: An overview of the bioinformatics software for metabarcoding data analyses. *Authorea. Mol. Ecol. Res.*, 1–17.

Hallmann, C.A., Sorg, M., Jongejans, E., Siepel, H., Hofland, N., Schwan, H., Stenmans, W., Muller, A., Sumser, H., Horren, T. et al. (2017). More than 75 percent decline over 27 years in total flying insect biomass in protected areas. *PLOS One*, 12(10), e0185809.

Hartop, E., Srivathsan, A., Ronquist, F., Meier, R. (2022). Towards large-scale integrative taxonomy (LIT): Resolving the data conundrum for dark taxa. *Syst. Biol.*, 71(6), 1404–1422.

Hebert, P.D.N., Cywinska, A., Ball, S.L., Dewaard, J.R. (2003). Biological identifications through DNA barcodes. *P. Roy. Soc. B-Biol. Sci.*, 270, 313–321.

Hebert, P.D.N., Braukmann, T.W.A., Prosser, S.W.J., Ratnasingham, S., Dewaard, J.R., Ivanova, N.V., Janzen, D.H., Hallwachs, W., Naik, S., Sones, J.E. et al. (2018). A sequel to Sanger: Amplicon sequencing that scales. *BMC Genom.*, 19(1), 219.

Hendrich, L., Morinière, J., Haszprunar, G., Hebert, P.D., Hausmann, A., Kohler, F., Balke, M. (2014). A comprehensive DNA barcode database for Central European beetles with a focus on Germany: Adding more than 3,500 identified species to BOLD. *Mol. Ecol. Resour.*, 15(4), 795–818.

IPBES (2019). Global assessment report on biodiversity and ecosystem services of the intergovernmental science-policy platform on biodiversity and ecosystem services. Report, IPBES Secretariat, Bonn.

Janzen, D.H. and Hallwachs, W. (2021). To us insectometers, it is clear that insect decline in our Costa Rican tropics is real, so let's be kind to the survivors. *Proc. Natl. Acad. Sci. USA*, 118(2).

Kaczmarek, M., Entling, M.H., Hoffmann, C. (2022). Using malaise traps and metabarcoding for biodiversity assessment in vineyards: Effects of weather and trapping effort. *Insects*, 13(6), 507.

Kerr, K.C.R., Stoeckle, M.Y., Dove, C.J., Weigt, L.A., Francis, C.M., Hebert, P.D.N. (2007). Comprehensive DNA barcode coverage of North American birds. *Mol. Ecol. Notes*, 7(4), 535–543.

Kirse, A., Bourlat, S.J., Langen, K., Fonseca, V.G. (2021). Metabarcoding malaise traps and soil eDNA reveals seasonal and local arthropod diversity shifts. *Sci. Rep.*, 11(1), 10498.

Lannom, L., Koureas, D., Hardisty, A.R. (2020). FAIR data and services in biodiversity science and geoscience. *Data Intelligence*, 2(1/2), 122–130.

Le Gall, L., Delsuc, F., Hourdez, S., Lecointre, G., Rasplus, J.-Y. (2017). Toward the DNA library of life. *Eur. J. Taxon.*, 266.

Lees, D.C., Kawahara, A.Y., Rougerie, R., Oshima, I., Kawakita, A., Bouteleux, O., De Prins, J., Lopez-Vaamonde, C. (2014). DNA barcoding reveals a largely unknown fauna of Gracillariidae leaf mining moths in the Neotropics. *Mol. Ecol. Resour.*, 14(2), 286–296.

Lévêque, A. and Witté, I. (2019). Les enjeux de biodiversité en France métropolitaine : analyses croisées. Report, Ministère de la Transition Écologique et Solidaire, Paris.

Liu, C., Ashfaq, M., Yin, Y., Zhu, Y., Wang, Z., Cheng, H., Hebert, P. (2023). Using DNA metabarcoding to assess insect diversity in citrus orchards. *Peer J.*, 11.

Lopez-Vaamonde, C., Kirichenko, N., Cama, A., Doorenweerd, C., Godfray, H.C.J., Guiguet, A., Gomboc, S., Huemer, P., Landry, J.-F., Laštůvka, A. et al. (2021). Evaluating DNA barcoding for species identification and discovery in European Gracillariid Moths. *Front Ecol. Evol.*, 9.

Meier, R., Wong, W., Srivathsan, A., Foo, M. (2015). $1 DNA barcodes for reconstructing complex phenomes and finding rare species in specimen-rich samples. *Cladistics*, 32(1), 100–110.

Morinière, J., Cancian De Araujo, B., Lam, A.W., Hausmann, A., Balke, M., Schmidt, S., Hendrich, L., Doczkal, D., Fartmann, B., Arvidsson, S. et al. (2016). Species identification in malaise trap samples by DNA barcoding based on NGS technologies and a scoring matrix. *PLOS One*, 11(5), e0155497.

Moritz, C. and Cicero, C. (2004). DNA barcoding: Promise and pitfalls. *PLOS Biol.*, 2(10), e354.

Padial, J.M., Miralles, A., De La Riva, I., Vences, M. (2010). The integrative future of taxonomy. *Front Zool*, 7, 16.

Pomerantz, A., Penafiel, N., Arteaga, A., Bustamante, L., Pichardo, F., Coloma, L.A., Barrio-Amoros, C.L., Salazar-Valenzuela, D., Prost, S. (2018). Real-time DNA barcoding in a rainforest using nanopore sequencing: Opportunities for rapid biodiversity assessments and local capacity building. *Gigascience*, 7(4), giy033.

Porter, T.M. and Hajibabaei, M. (2018). Scaling up: A guide to high-throughput genomic approaches for biodiversity analysis. *Mol. Ecol.*, 27(2), 313–338.

Ratnasingham, S. and Hebert, P.D.N. (2007). BOLD: The barcode of life data system. *Mol. Ecol. Notes*, 7(3), 355–364.

Ratnasingham, S. and Hebert, P.D.N. (2013). A DNA-based registry for all animal species: The Barcode Index Number (BIN) System. *PLOS ONE*, 8(7), e66213.

Rimet, F., Aylagas, E., Borja, A., Bouchez, A., Canino, A., Chauvin, C., Chonova, T., Ciampor Jr, F., Costa, F.O., Ferrari, B.J.D. et al. (2021). Metadata standards and practical guidelines for specimen and DNA curation when building barcode reference libraries for aquatic life. *Metabarcoding Metagenom*, 5, e58056.

Royaux, C., Arnaud, E., Sananikone, J., Jossé, M., Madelin, M., Pelletier, D., Norvez, O., Le Bras, Y. (2022). Open science for better FAIRness: A biodiversity virtual research environment point of view. *Biodivers. Inf. Sci. Stand*, 6, e95110.

Rubinoff, D. and Holland, B.S. (2005). Between two extremes: Mitochondrial DNA is neither the panacea nor the nemesis of phylogenetic and taxonomic inference. *Syst. Biol.*, 54(6), 952–961.

Sanger, F., Nicklen, S., Coulson, A.R. (1977). DNA sequencing with chain-terminating inhibitors. *Proc. Natl. Acad. Sci. USA*, 74(12), 5463–5467.

Santos, B., Miller, M., Miklasevskaja, M., Mckeown, J., Redmond, N., Coddington, J., Bird, J., Miller, S., Smith, A., Brady, S. et al. (2023). Enhancing DNA barcode reference libraries by harvesting terrestrial arthropods at the Smithsonian's National Museum of Natural History. *Biodivers. Data J.*, 11.

Schoch, C.L., Seifert, K.A., Huhndorf, S., Robert, V., Spouge, J.L., Levesque, C.A., Chen, W., Fungal Barcoding Consortium, Fungal Barcoding Consortium Author List (2012). Nuclear ribosomal internal transcribed spacer (ITS) region as a universal DNA barcode marker for Fungi. *Proc. Natl. Acad. Sci. USA*, 109(16), 6241–6246.

Setsuko, S., Yoshimura, K., Ueno, S., Worth, J.R.P., Ujino-Ihara, T., Katsuki, T., Noshiro, S., Fujii, T., Arai, T., Yoshimaru, H. (2023). A DNA barcode reference library for the native woody seed plants of Japan. *Mol. Ecol. Resour.*, 23(4), 855–871.

Shokralla, S., Porter, T.M., Gibson, J.F., Dobosz, R., Janzen, D.H., Hallwachs, W., Golding, G.B., Hajibabaei, M. (2015). Massively parallel multiplex DNA sequencing for specimen identification using an Illumina MiSeq platform. *Sci. Rep.*, 5, 9687.

Sire, L., Yanez, P.S., Wang, C., Bezier, A., Courtial, B., Cours, J., Fontaneto, D., Larrieu, L., Bouget, C., Thorn, S. et al. (2022). Climate-induced forest dieback drives compositional changes in insect communities that are more pronounced for rare species. *Commun. Biol.*, 5(1), 57.

Srivathsan, A., Lee, L., Katoh, K., Hartop, E., Kutty, S.N., Wong, J., Yeo, D., Meier, R. (2021). ONTbarcoder and MinION barcodes aid biodiversity discovery and identification by everyone, for everyone. *BMC Biol.*, 19(1), 217.

Stork, N.E. (2018). How many species of insects and other terrestrial arthropods are there on Earth? *Annu. Rev. Entom.*, 63, 31–45.

Taberlet, P., Coissac, E., Pompanon, F., Brochmann, C., Willerslev, E. (2012). Towards next-generation biodiversity assessment using DNA metabarcoding. *Mol. Ecol.*, 21, 2045–2050.

Touroult, J., Chaumet, S., Poncet, L., Siblet, J.-P. (2017a). Diagnostic et recommandations pour une stratégie d'acquisition de connaissances naturalistes continentales. Tome I : analyse des besoins et des dispositifs existants. Report, MNHN, Paris.

Touroult, J., Chaumet, S., Poncet, L., Siblet, J.-P. (2017b). Diagnostic et recommandations pour une stratégie d'acquisition de connaissances naturalistes continentales. Tome II : synthèse et propositions d'actions. Report, MNHN, Paris.

Wägele, J.W., Bodesheim, P., Bourlat, S.J., Denzler, J., Diepenbroek, M., Fonseca, V., Frommolt, K.-H., Geiger, M.F., Gemeinholzer, B., Glöckner, F.O. et al. (2022). Towards a multisensor station for automated biodiversity monitoring. *Basic Appl. Ecol.*, 59, 105–138.

Wagner, D.L., Grames, E.M., Forister, M.L., Berenbaum, M.R., Stopak, D. (2021). Insect decline in the Anthropocene: Death by a thousand cuts. *Proc. Natl. Acad. Sci. USA*, 118(2), e2023989118.

Weigand, H., Beermann, A.J., Ciampor, F., Costa, F.O., Csabai, Z., Duarte, S., Geiger, M.F., Grabowski, M., Rimet, F., Rulik, B. et al. (2019). DNA barcode reference libraries for the monitoring of aquatic biota in Europe: Gap-analysis and recommendations for future work. *Sci. Total Environ.*, 678, 499–524.

West, K., Travers, M.J., Stat, M., Harvey, E.S., Richards, Z.T., Dibattista, J.D., Newman, S.J., Harry, A., Skepper, C.L., Heydenrych, M. et al. (2021). Large-scale eDNA metabarcoding survey reveals marine biogeographic break and transitions over tropical north-western Australia. *Divers. Distrib.*, 27(10), 1942–1957.

Wilkinson, M.D., Dumontier, M., Aalbersberg, I.J., Appleton, G., Axton, M., Baak, A., Blomberg, N., Boiten, J.W., Da Silva Santos, L.B., Bourne, P.E. et al. (2016). The FAIR guiding principles for scientific data management and stewardship. *Sci. Data*, 3, 160018.

Will, K.W., Mishler, B.D., Wheeler, Q.D. (2005). The perils of DNA barcoding and the need for integrative taxonomy. *Syst. Biol.*, 54(5), 844–851.

Young, M.R. and Hebert, P.D.N. (2022). Unearthing soil arthropod diversity through DNA metabarcoding. *Peer J.*, 10, e12845.

7

Exploring the Molecular Biodiversity of Specimens in Collections: The Case of Coccinellidae

7.1. Introduction

Museums, herbariums, and biological collection archives (for example, document archives, sound archives, engravings, photographs, maps, etc.) constitute natural history collections (NHCs) (Wandeler et al. 2007). NHCs are considered spatiotemporal testimonials to biological diversity, providing information as to the presence of species in a given place and at a given time. Therefore, NHCs are remarkable assets which help us to understand the origin and evolution of biodiversity (Meredith 1996; Shaffer et al. 1998; Nattier 2018), but they remain under-used in comparison to their true richness, notably in terms of their genetic resources (Wandeler et al. 2007; Yeates et al. 2016).

Museomics is a growing approach that exploits the potential of specimens in natural history museums as a source of DNA. Thanks to progress made in high-throughput sequencing technologies, also known as *Next-Generation Sequencing* (NGS) (Goodwin et al. 2016), the use of museum collections as sources of genetic data has increased. This extended use of museum specimens has further increased their value and allowed them to be included in new interdisciplinary research questions (conservation biology; evolutionary biology: genetics of populations, phylogenetics, phylogeography; ecology, environmental changes, and resource management, etc.), and

---

Chapter written by Karen SALAZAR, Romain NATTIER and Guillaume ACHAZ.

finally, reaffirmed their roles (Meredith 1996; Lopez et al. 2020; Benham and Bowie 2023). The integration of traditional museum practices and genomics leads us to envisage new practices for the conservation and optimization of the use of genetic sources in museums (Card et al. 2021).

In this chapter, we present the general context of the use of insect specimens from NHCs to open the door to the study of their diversity and their evolution through historical DNA sequencing. We also present an overview of the most widely used DNA sequencing methods in these fields and the recent results of museomics in entomology. As a case study, we summarize the results obtained in this field concerning Coccinellidae (Coleoptera), which is the biological model of interest to us.

7.2. DNA sequencing of collection specimens

Museomics – a portmanteau of *museum* and *genomics* – refers to the study of DNA obtained from specimens or samples from NHCs (Raxworthy and Smith 2021) preserved traditionally for at least 200 years. This genetic material is called historical DNA (hDNA), or quite simply, DNA taken from museum samples. It differs from ancient DNA (aDNA) in that the latter is obtained from samples (hard tissues or environmental samples) naturally preserved for centuries, millennia, or even millions of years at paleontological and archaeological sites (Wandeler et al. 2007; Raxworthy and Smith 2021). Every specimen in a museum's collection bears unique importance and is a primary source of valuable information. Some information, such as data on morphology, and genetic and molecular makeup (e.g. chemical components), may be pertinent. Other information may include data linked to the habitat, ecology, place in time, and geographic location, including data about symbiotes (e.g. gut microbiome, and pathogenic agents), which, in turn, will have associated data (Miller et al. 2020). These data are at the heart of numerous domains of scientific research. Museomics has significant benefits when we do not have fresh or alcohol-preserved samples, and when missions on the ground for new collections are difficult or even impossible to carry out (due to inaccessible habitats, geopolitical problems, etc.). These difficulties may be mitigated by using museum collections, which offer the possibility of obtaining a large number of samples, spread out across time and space.

7.3. Methodology for DNA sequencing of collection specimens

7.3.1. *Extraction of DNA from collection specimens*

The protocols for extracting DNA without destroying the specimens or samples (Gilbert et al. 2007; Thomsen et al. 2009; Tin et al. 2014) have helped to increase the number of specimens included in museomic studies, such as nomenclatural types and unique or rare samples, as the DNA can be extracted from the specimen without visible external morphological damage. This means the specimen can be returned to the collection after the DNA extraction process and can be used for other studies. Extraction protocols vary, depending on the target group of insects. Generally, dry-preserved specimens are cleaned, taken off their pin, or detached from their label, and then placed wholly into a digestion buffer, following the protocol set forth by Gilbert et al. (2007). It is also possible to use part of the specimen's body (the feet, the abdomen, the head) without destroying it, for DNA extraction. Then, the specimens or samples are washed with double-distilled water to eliminate all traces of extraction buffer, dehydrated, air-dried, reassembled, and returned to the original collection. This DNA extraction method has a minimal impact on the specimen's external morphology, except for changes in pigmentation – especially in species that have particularly vibrant colors (which is the case with the majority of ladybirds), and sometimes certain tarsi and antennae, which may become detached from the specimen. A study has been conducted on the effect of different pretreatments for DNA extraction from dried specimens of four species of Coccinellidae collected between 2008 and 2012, demonstrating that 0.9% NaCl buffer and STE buffer produce the highest quality and purity of DNA (Huang et al. 2019).

Liquid nitrogen, ethanol at a concentration of over 70%, supersaturated saline solution (Nagy 2010), 99% propylene glycol (Nakahama et al. 2019), and low-temperature freezing of "cryogenic collections" (Card et al. 2021) help to preserve DNA over time, in freshly sampled specimens/tissues.

7.3.2. *Difficulties in the molecular use of collection specimens*

The main difficulty in the molecular use of collection specimens is a technical obstacle. The degradation and fragmentation of specimens' DNA, post-mortem, and after they are added to a collection, reduce the quality and quantity of usable DNA in the oldest specimens. Nonetheless, new technologies for high-throughput sequencing are capable of sequencing highly fragmented DNA, and more fully exploiting the genetic resources of NHCs. There are different types of molecular damage that aDNA may suffer, such as strand breaks, breaks in the transverse connections in the DNA, hydrolytic lesions; or contamination by exogenous DNA (Pääbo et al. 2004), which is also present in the hDNA of insect specimens from museums (Mullin et al. 2023). The success of sequencing from museum specimens depends on multiple factors, or a combination of factors. The first such factor is the age of the specimens – though it is not a directly linked factor, their size, their state of conservation – for example, if they have been exposed to pathogenic agents, the type of fixing used for their display, and how they have been stored before being mounted. The conservation conditions in which the collection has been kept (light levels, humidity, temperature, chemical treatment used to treat harmful insects in the NHC) also need to be taken into consideration, as do the quality and quantity of DNA obtained, how it is handled in the laboratory, and the sequencing strategy chosen (e.g. Espeland et al. 2010; Kanda et al. 2015; Mayer et al. 2021; and Nunes et al. 2022).

7.3.3. *Next-Generation Sequencing (NGS) technologies used in museomics*

Though there are many different platforms for NGS, Illumina machines are the most popular; they can be used for sequencing historic and ancient DNA that is badly degraded and fragmented. Concerning insects, the majority of studies using NGS have been performed with Illumina platforms (> 60%) (Matos-Maraví et al. 2019). Among the approaches that can be used to sequence the DNA obtained from museum specimens or samples (Rizzi et al. 2012; Burrell et al. 2015; Yeates et al. 2016; Raxworthy and Smith 2021), we can point to the following as particular examples: a) whole-genome shotgun (*WGS*) sequencing, with little coverage, also known as "genome skimming". This method enables us, randomly, to obtain a fraction of the genome with numerous copies, such as mitochondria and chloroplasts, ribosomes, and even certain genes with a small number of copies but numerous samples (Straub et al. 2012) (e.g. insects from museum collections (Zhou et al. 2022)); b) restriction-site-associated DNA sequencing (*RADseq.*), which consists of using restriction enzymes to select and sequence regions throughout the genome (Tin et al. 2014), and *RAD Capture*, an alternative RAD protocol (Ali et al. 2016); c) hybridization capture using RAD probes (*HyRAD*) is a sequencing method which uses

RAD probes on fresh specimens with high-quality DNA as "bait" to capture homologous fragments of interest from genome libraries of museum specimens (Suchan et al. 2016), or HyRAD-X, which uses mRNA from fresh specimens as a model to design probes based on RAD sequencing (double digest RAD), which are then used to capture homologous DNA from other samples (Schmid et al. 2017) (e.g. see Toussaint et al. 2021; Gauthier et al. 2023); and d) Sequence-capture, also known as *Hybrid enrichment*, which entails selectively enriching specific regions of a genome using probes (Lemmon et al. 2012) (e.g. see Blaimer et al. 2016; St Laurent et al. 2018; Mayer et al. 2021; Twort et al. 2021)).

7.4. Recent results from museomic studies on insects

Since 2005, with the introduction and progress of NGS platforms (Goodwin et al. 2016), the number of projects aiming to collect genetic data from museum specimens has grown considerably. These technological advances, the increase in gainfulness of such studies (in terms of the number of samples and data sequenced), and the ease of obtaining a large number of sequenced loci (Goodwin et al. 2016; Linderholm 2016) have extended the list of taxa which can be studied using NGS technologies.

However, the use of insects from museums for DNA studies began before the introduction of NGS and was based on Sanger sequencing (for example, see Townson et al. 1999; Goldstein and Desalle 2003), which continues to be used today (see Table 7.1). In the specific case of insects, DNA sequencing of museum collection specimens goes back to 2006. Here, we can cite the pioneering studies of Harper et al. (2006) and Hartley et al. (2006), who measured the genetic changes between populations, by comparing museum samples to recent samples. These studies raise important questions on the subject of reduction in allele frequency caused by habitat fragmentation and increase in insecticide resistance over time.

Figure 7.1, starting from 2006, shows a progressive increase in the number of publications using historical DNA of insects. From 2018, this number has increased very substantially, reaching 100 publications in all in 2023, notably thanks to the development of NGS technologies. These surprising and very promising results (see both Figure 7.1 and Table 7.1) are cited here, to highlight the effectiveness of these new approaches to DNA sequencing using museum insect specimens.

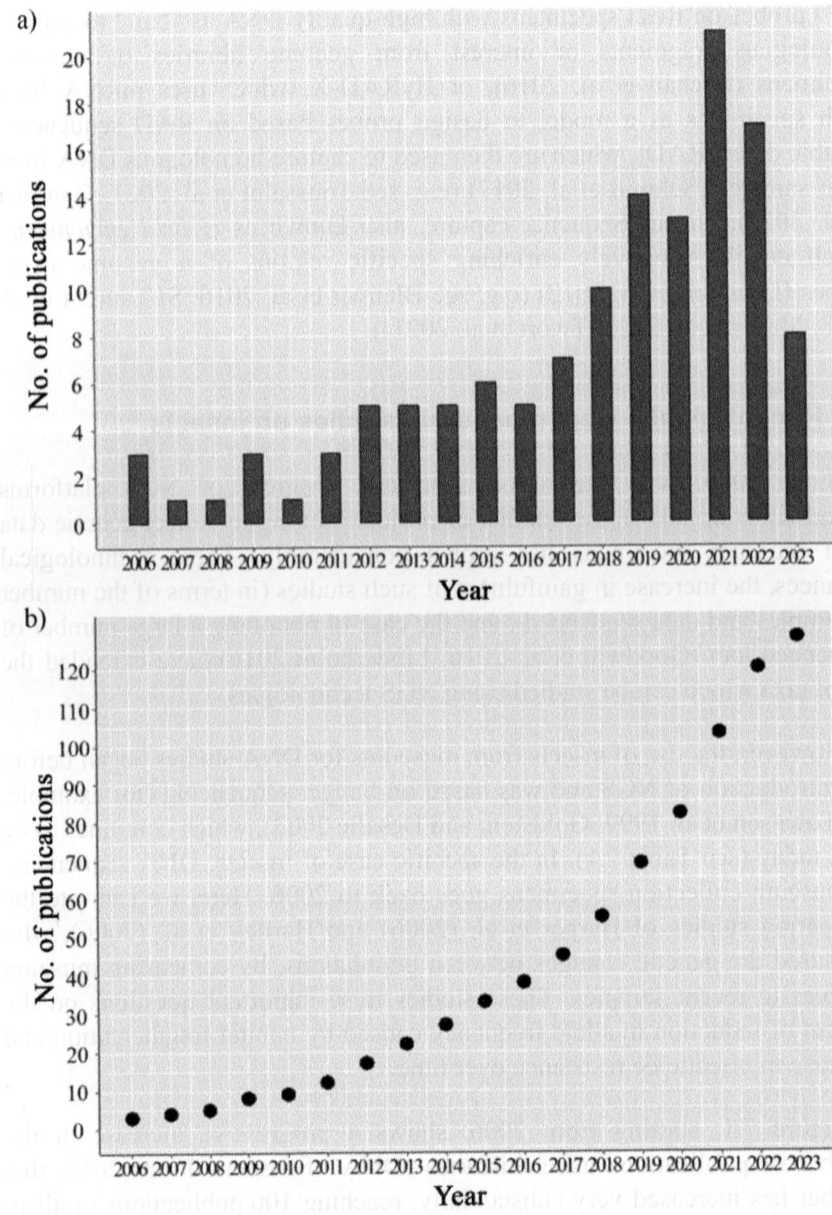

Figure 7.1. *Number of articles published using historical DNA from insects preserved in natural history collections over time, following the introduction of NGS technologies in 2005. a) Number of publications per year, and b) cumulative number of publications by year*

COMMENT ON FIGURE 7.1.– *Based on a Google Scholar search using the search terms: "historical DNA, museomics, historical specimens, old museum specimens, museum genomics, NGS, insects"; the data include a range of DNA sequencing approaches and the analysis does not include theses or reviews. The search was performed in June 2023.*

Table 7.1 gives a few examples of historical DNA sequenced from most ancient insects conserved in museums, including specimens collected from the late 1700s and the early 1800s and 1900s.

The acquisition of a relatively high number of genes from museum specimens met with great success. It has grown from small fragments of certain genes, including the most repetitive fragments in the nuclear genome (ribosomes) and mitochondrial genes such as the gene for Cytochrome c oxidase I (COI), to entire mitogenomes, numerous nuclear genes, partial genomes, and nearly complete genomes. These results have come from various sequencing techniques using diverse NGS technologies.

These studies suggest that, in the near future, it should be possible to sequence insects from the historical collections of famous naturalists from the 18th and 19th Centuries – for example, insects collected by C. Darwin (1831–1836) and named collections such as those of Lamarck (1744–1829), Olivier (1756–1814), Latreille (1762–1833), Dejean (1780–1845), Germar (1786–1853), Mulsant (1797–1880), Bates (1825–1892), and Wallace (1823–1913). For example, the Muséum national d'Histoire naturelle (MNHN), in Paris houses a vast number of specimens collected by naturalists from past centuries (Cambefort 2006).

It is estimated that around 1–3 billion specimens are preserved in NHCs (Ariño 2010; Yeates et al. 2016). In total, the MNHN's collections contain about 68 million specimens, and specifically, the entomology collection contains around 20 million specimens of Coleoptera, including some 185,000 type specimens, which represent 25% of total collections and 50% of insect collections[1].

1 Available at: www.mnhn.fr [Accessed June 2023].

Order/Family	Taxon	Age (years of collection)	Approach	Genomic data obtained	Reference
Coleoptera					
Byrrhidae	*Byrrhus pilula* (Linnaeus)	(1820–2005)	SS	COI fragments	(Thomsen et al. 2009)
Carabidae	*Amara alpina* (Paykull)	~ 136–7 (1875–2004)	SS; WGS*	COI, 28S fragments	(Heintzman et al. 2014)
	Bembidion Latreille	69-recent (1945–2012)	WGS*	COI, 18S, 28S, up to 67 fragments of nuclear PCGs with low-copy-number	(Kanda et al. 2015)
	Species of *Bembidion*	~ 136–77 (1878–1939)	WGS*	COI, 28S, 18S, CAD, Topo I fragments	(Sproul and Maddison 2017)
	Bembidion lividulum Casey	100	WGS	rDNA	(Sproul and Maddison 2018)
	Harpalus latus (Linnaeus)	(1825–2004)	SS	COI fragments	(Thomsen et al. 2009)
	Lionepha sp.	58 (1956)	WGS*	COI, 18S, 28S, up to fragments of nuclear PCGs with low-copy-number	(Kanda et al. 2015)
	Species of *Lionepha* Casey	~ 159–58 (1853–1958)	WGS*	COI, 28S, 18S, CAD, Topo I fragments	(Sproul and Maddison 2017)
	Species of Pausinae	52–2 (1952–2002)	SS	COI, 28S fragments	(Gilbert et al. 2007)
Cerambycidae	Many species	128–2	WGS	Mitochondrial genes	(Jin et al. 2020)

Order/Family	Taxon	Age (years of collection)	Approach	Genomic data obtained	Reference
Chrysomelidae	*Chrysolina polita* (Linnæus)	(1899)	SS	COI fragments	(Thomsen et al. 2009)
	28 species of *Oreina* Chevrolat and another 19 leaf beetles	(1898–2017)	HyRAD	2,235 nuclear loci	(Gauthier et al. 2023)
Coccinellidae	*Eriopis patagonia* Salazar	~ 100 (< 1930)	WGS*	Mitogenome	(Salazar and Nattier 2020)
Curculionidae	*Otiorhynchus sulcatus* (Fabricius)	(1884–1998)	SS	COI fragments	(Thomsen et al. 2009)
	Curculio pyrrhoceras Marsham	(1896–2000)	SS	COI fragments	(Thomsen et al. 2009)
Dermestidae	Multiple species	106–2 (< 1905–2015)	WGS	Mitogenome	(Zhou et al. 2022)
Tenebrionidae	Lagriinae sp.	84 (1929)	WGS*	COI, 18S, 28S, 25 fragments of nuclear PCGs with low-copy-number	(Kanda et al. 2015)
		Diptera			
Agromyzidae	*Japanagromyza tokunagai* Sasakawa	(1923–2016)	SS	COI fragments	(Yamashita et al. 2022)
Drosophilidae	*Drosophila melanogaster* Meigen	(~ 1810–1933)	WGS	Nearly complete nuclear genomes	(Shpak et al. 2023)
8 families (Oestroidea)	98 species	92–2	SC	37–2,077 UCE loci	(Buenaventura et al. 2021)

Order/Family	Taxon	Age (years of collection)	Approach	Genomic data obtained	Reference
		Hymenoptera			
Apidae	*Apis mellifera* Linnaeus	141–61 (1879–1959)	WGS	Mitogenome, 664 nuclear genes	(Parejo et al. 2020)
	Xylocopa Latreille	121–2 (1894–2013)	SC	6–972 nuclear loci	(Blaimer et al. 2016)
Bombidae	*Bombus lapidarius* (Linnaeus)	(1891–2004)	WGS	0.01–96% coverage 3x of the nuclear genome	(Mullin et al. 2023)
Braconidae	*Atanycolus longifemoralis* Shenefelt	70–29 (1940–1981)	SS	COI, 28S fragments	(Andersen and Mills 2012)
Formicidae	4 species	(1910–1953)	RAD-Seq	A few SNPs	(Tin et al. 2014)
		Lepidoptera			
Epicopeiidae	17 species	(1906–2001)	SC	468–517 CDS	(Call et al. 2021)
Hesperiidae	*Hesperia colorado* (Scudder)	(1871–2017)	WGS	Mitogenome, partial nuclear genomes	(Cong et al. 2021)
Hesperidae	36 species of *Pyrgus* Hübner	(1929–2012)	WGS*	Mitogenome, nuclear rRNA operon	(Pitteloud et al. 2017)
	Pyrrhopyginae	~ 140-recent (1885–2016)	SS WGS	COI fragments Mitogenomes, nuclear PCGs	(Zhang et al. 2019)
Lycaenidae	*Glaucopsyche xerces* (Boisduval)	93 (1928)	WGS*	Mitogenome	(Grewe et al. 2021)
	Lycaena helle (Denis and Schiffermüller)	(1957–2007)	HyRAD	High number of loci	(Suchan et al. 2016)
	Polyommatus bellargus (Rottemburg)	100 (1896)	SS	5 microsatellites	(Harper et al. 2006)

Order/Family	Taxon	Age (years of collection)	Approach	Genomic data obtained	Reference
Sphingidae	17 species of *Hyles* Hübner	241–3 (1779–2017)	WGS	Mitogenome	(Patzold et al. 2020)
	Hyles calida hawaiiensis Rothschild and Jordan	100 (~ 1907)	SS	COI, COII fragments	(Hundsdoerfer and Kitching 2017)
27 families	33 species	(1892–2017)	SC	COI, up to 2,953 loci	(Mayer et al. 2021)
6 families (Papilionoidea)	Multiple species	111–0	SC	Set of 391 loci or 13 loci	(Nunes et al. 2022)
Odonata					
19 families	Multiple species	~ 112–20 (1909–2001)	SC	4–1,049 loci	(Goodman et al. 2023)
Neuroptera					
Chrysopidae	*Chrysoperla carnea* (Stephens)	180	SS	COI fragments	(Price et al. 2015)
Orthoptera					
Acrididae	*Oedaleus decorus* (Germar)	(1908–1997)	HyRAD	High number of loci	(Suchan et al. 2016)
Tettigoniidae	*Neoconocephalus lyristes* Rehn and Hebard	(1905–1932)	SS	COI fragments	(Ney et al. 2018)

Table 7.1. *Sequencing of historical DNA from some of the most ancient insect specimens found in museum collections. Age – years before DNA extraction (collection date of the specimen). Sequencing approach – HyRAD: Hybridization RAD; SC: sequence capture; SS: Sanger sequencing; WGS: Whole Genome Shotgun, * indicates genome skimming. Coding sequence (CDS); protein-coding gene (PCG); single-nucleotide polymorphisms (SNPs); Ultra-Conserved Element (UCE)*

The study of historical DNA from insect specimens in museums has helped put forward new phylogenies (e.g. see Van Dam et al. 2018; Johnston et al. 2022; Zhou et al. 2022), assess the phylogenetic relations of extinct lineages (e.g. see De-Dios et al. 2023), accurately identify species using the DNA of type specimens (e.g. see Price et al. 2015; Sproul and Maddison 2017; Patzold et al. 2020), reveal cryptic species and speed up the process of species discovery (for example, see Sproul and Maddison 2018; Jin et al. 2020), study genetic changes to populations over time in the Anthropocene (e.g. Mikheyev et al. 2015, 2017; Gauthier et al. 2020; Brasil et al. 2023), and measuring climate change (Ryan et al. 2018). These collections also enable us to study interactions with pathogens and symbiotes (e.g. Tay et al. 2017), obtain information on the microbiome (e.g. Linard et al. 2016; Yamashita et al. 2022), and constitute archives of chemical information (e.g. Berger et al. 2021; Moris et al. 2021), to cite only a few research areas.

7.5. Context of the study on biodiversity and systematics of Coccinellidae

Ladybirds (Coccinellidae), with their vibrant colors, are insects that the general public can readily recognize. They are endearing and emblematic, and have important meanings in many cultures (Majerus 2016). Most species of Coccinellidae are predators of aphids and other Hemiptera such as scale insects, whiteflies, psyllids, and other small arthropods, while others are phytophages (for example, eating plant tissues and pollen), feed on fungi (spores, mold), or have a mixed diet (Escalona et al. 2017; Nattier et al. 2021). Given their predatory behavior, they are used to keep pests at bay and play an important role in the regulation of agroecosystems, orchards, and gardens (Majerus 2016). This family includes around 6,000 species that have been described, all over the world, classified into 360 genera and three sub-families (Coccinellinae, Mircroweiseinae, and Monocoryninae) (Ślipiński 2007; Che et al. 2021). Coccinellidae are a very widely studied group, because they are highly diversified, have short life cycles, with several generations per year, and have a wide distribution, occupying a variety of habitats the world over.

Recent progress has been made in knowledge of the phylogenetic relations of the family. The following are the four most recent large-scale molecular phylogenies with estimation of divergence time. Li et al. (2020) published a phylogeny of the subfamily Chilocorini, comprising 22 genera, and 82 species, using five protein-coding genes (PCGs). The Coccinellini subfamily, comprising 90 genera and over 1,000 species, was the subject of two studies in 2021: one by Tomaszewska et al. 2021 (4,214 nucleotides of DNA sequences, 6 PCGs from 164 species) and the other by Nattier et al. 2021 (8,158 nucleotides of DNA sequences, 8 PCGs from 105 species). In these results, the authors demonstrated the paraphyly of multiple genera and introduced changes to the nomenclature. A broader phylogeny (77,000 nucleotides of DNA sequences, 94 PCGs from 214 species) of the family was produced by Che et al. (2021), which allowed more robust hypotheses to be put forward on the phylogenetic relations between the different subfamilies and the different genera, the divergence periods for the main lineages and the bases for more accurate nomenclatural classification of the taxa of Coccinellidae. Although these phylogenies do not include collection specimens, the use of museomics will soon become essential in studies of this family.

7.5.1. *Applications of museomics for exploring the biodiversity of Coccinellidae*

Molecular studies of Coccinellidae using historical DNA are rare currently, but the potential use of museum specimens in this group will increase rapidly in years to come. Using a non-destructive extraction protocol and NGS on an Illumina platform, we have been able to sequence the DNA of numerous historic specimens from collections (see Figure 7.2). The first such project was the sequencing of the complete mitochondrial genome of *Coccidophilus cariba* Gordon, 1978, from a specimen collected in Guadeloupe 22 years previously, housed in the MNHN's collection in Paris, revealing its length to be 15,343 base pairs (bp). This was the first ever mitochondrial genome to be published for the subfamily Microweiseinae and the genus *Coccidophilus* Brethes (Nattier and Salazar 2019). Secondly, *Eriopis patagonia* Salazar, 2020 was described based on two museum specimens collected some 100 years before in Patagonia, and deposited in the MNHN's collection. Using these two specimens, we described the mitochondrial genome of 16,194 bp, marking the first-ever description for this genus, and confirmed the phylogenetic position of the species in a phylogeny based on the mitochondrial genomes (Salazar and Nattier 2020).

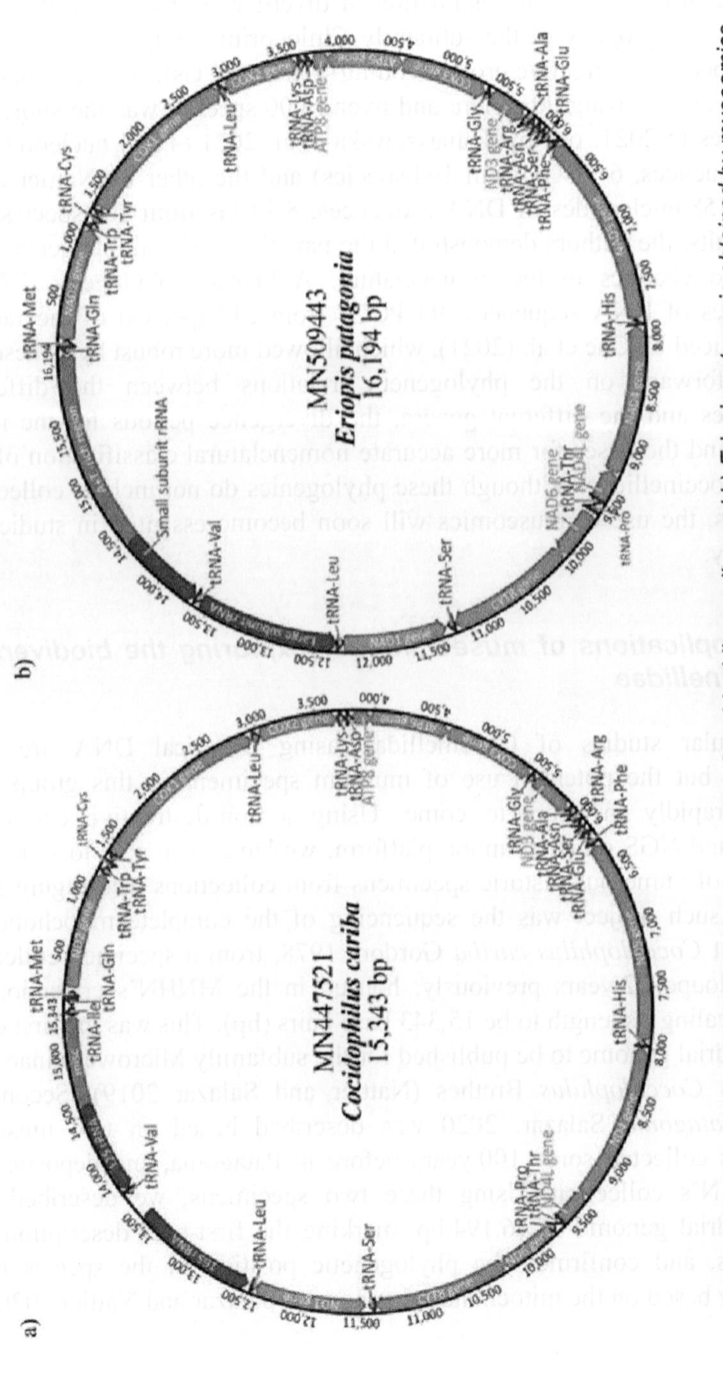

Figure 7.2. *Map of the mitochondrial genome of a) Coccidophilus cariba and b) Eriopis patagonia obtained by museomics from museum specimens. For a color version of this figure, see www.iste.co.uk/nicolas/biodiversity.zip*

COMMENT ON FIGURE 7.2.– *The 13 protein-coding genes (PCGs) are shown in green, the 21 transfer RNAs (tRNA) in blue, the two ribosomal RNAs (rRNA) in red, and the region rich in A+T in orange. The transcription direction is shown by an arrow running along the mitogenome.*

As part of ongoing projects on ladybird beetles using museomics, we sequenced nearly 200 specimens from different NHCs, representing numerous taxonomic samples (around 60 genera) across time and space. The age of the samples varied between 1 and 186 years, from the date of the specimen collection to the date of DNA extraction. In the first project, we focused on the South American genus *Eriopis* Mulsant, 1850, comprising 24 species, mainly present in the Andes. We studied only collection specimens, explored different morphological traits, and distribution data, and extracted DNA from 59 specimens representing 16 species. The genomic data were obtained by NGS, mainly comprised mitochondrial genomes, the most repetitive genes from the nuclear genome as the ribosomal operon and histone complex, and some genes with low-copy-number. With an integrative approach, we have shed light on the systematics, the origin, and the process of diversification of this genus.

In a second project, we focused on the genera *Coccinella* Linnaeus 1758, and *Lioadalia* Crotch, 1874, which are very similar from a morphological point of view. *Coccinella*, with 44 species, is cosmopolitan, but the majority are found in the Nearctic and Palearctic realms. *Lioadalia* includes numerous Afrotropical species. In this study, we sampled 60 museum specimens collected over a period from 1966 to 2016, representing 67% of their diversity. The phylogeny was first deduced based on the 13 mitochondrial PCGs, and then an annotated genome of *Coccinella septempunctata* Linnaeus 1758 was used as a reference, yielding up to 145 coding sequences. The earliest phylogenetic analyses revealed that *Lioadalia* is an Afrotropical lineage of *Coccinella*. Our results suggest synonymy of *Lioadalia* and expanding the distribution of the genus *Coccinella* in that region, confirming the results of previous studies (Nattier et al. 2021; Tomaszewska et al. 2021).

7.5.2. *Barriers to the exploration of Coccinellidae biodiversity*

Of the factors that make it difficult to explore the biodiversity, systematics, and evolution of Coccinellidae, we could point to the lack of identification of species, the taxonomic confusion between the species linked to the interspecies and intraspecies morphological similarity, the lack of taxonomic expertise, and undersampling. For a better understanding of the evolutionary history of the family and more solid hypotheses, it is important to supplement the sampling in terms of species richness, geographic distribution, historic life traits (e.g., dietary habits, morphology), and genomic data. Thus, genetic samples from scientific collections would constitute an interesting alternative.

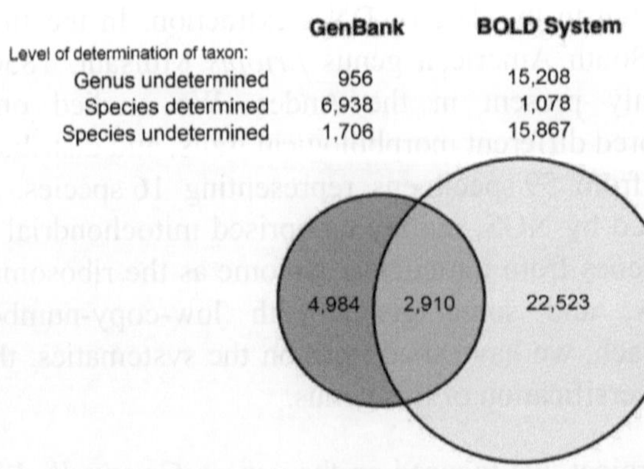

Figure 7.3. *Venn diagram showing the number of COI gene sequence records and the level of taxonomic determination in Coccinellidae from GenBank and BOLD databases. For a color version of this figure, see www.iste.co.uk/nicolas/biodiversity.zip*

Among the DNA barcodes that are most used for molecular identification of species is a sequence of 648 bp from the mitochondrial gene for COI (Ratnasingham and Hebert 2013). This DNA barcode has an appropriate genetic distance to allow it to be used in the identification of Coccinellidae (Huang et al. 2020). In the databases BOLD System and GenBank[2], there are 36,345 and 7,894 records of COI sequences for Coccinellidae, respectively (see Figure 7.3). In BOLD, only a few of these records have an assigned BIN (Barcode Index Number), 124 genera (40%) and 1,078 species (4%) are identified; whereas, for most of them, identification is still required (see Figure 7.3). In GenBank, the majority of records (87%) have species-level identification. These databases have 2,910 records in common (see

2 Available at: www.boldsystems.org and www.ncbi.nlm.nih.gov/genbank/ [Accessed June 2023].

Figure 7.3). The lack of taxonomic identification of Coccinellidae from BOLD and GenBank records reduces the use of these DNA sequences for precise studies on subjects like biodiversity, phylogenetics, and taxonomy. Museum specimens hold genetic information, including that which is used as DNA barcodes, and morphological, geolocation, and time-specific data, in addition to other data that are useful for integration into these two databases (Nakazato and Jinbo 2022) and to develop integrative projects in systematics, biogeography and evolution, among other research areas.

The availability of reference genomes is another limiting factor for genomic research on Coccinellidae. Advances in NGS technologies and rapid genome-annotation processes make studies on Coccinellidae much more feasible than they were, only a few years ago. To-date, various Coccinellidae nuclear genomes with different levels of assembling are available. In the database Resources of the National Center for Biotechnology Information (NCBI 1988), there are 22 nuclear genomes accessible for Coccinellidae, nine of which are at the level of chromosome clusters, six at the level of the contig, and seven at the level of the genome scaffolding. In addition, a genome has been filed in the National Genomics Data Center.[3]

Recently, annotated nuclear genomes of *Harmonia axyridis* (Pallas 1773) (size: 425.5Mega bases; 13,899 identified PCGs), *C. septempunctata* (size: 398.8Mb; 14,769 PCGs) (Crowley 2021), *Adalia bipunctata* (Linnaeus, 1758) (size: 475Mb; 13,611 PCGs) (Goate 2022), *Halyzia sedecimguttata* (Linnaeus, 1758) (size: 919.1Mb; 27,547 PCGs) (Crowley et al. 2023) and *Henosepilachna vigintioctomaculata* (Motschulsky, 1857) (size: 656.74Mb) (Zhu et al. 2023) are available. These annotated genomes will enable us to conduct studies in biology, adaptation, population genetics, gene expression, genetic bases of behavior, and other genetic, physiological, and phenotypic studies. They will also help shed light on the relations with endosymbionts, etc., in this family of Coleoptera. In addition, they will serve as a reference for more efficiently mapping and annotating genomic data obtained by genome skimming. Furthermore, they can be used as a basis for probe design in sequencing by hybrid enrichment.

3 See: www.ngdc.cncb.ac.cn.

7.6. Conclusion

The progress made in the exploitation of DNA from NHC specimens reflects the combined efforts of collectors, curators, taxonomists, developers of next-generation sequencing technologies and computational tools, and those who develop bio-informatic approaches to data analysis. NHCs contain billions of insect specimens that have the potential to be included in museomic studies, which further increases the value of each specimen. The study of historical DNA from insects is a rapidly expanding field, which allows for new and diverse applications in various research areas. Collections of fresh material in the field must continue to progress, and endure over the long term. These new collections will constitute the heritage of the future for researchers and society.

In the particular case of Coccinellidae, a better understanding of the evolutionary history and exploration of its diversity require a sampling effort that includes a greater richness, representing a broader range of life history traits (e.g., dietary habits and morphology), covering a larger number of geographic zones, to propose more solid phylogenetic and biogeographical hypotheses about the lineages. Museomics is an ideal strategy for pursuing and developing new research in this family of Coleoptera. The genetic data obtained in the different projects must also be integrated into databases containing molecular barcodes, to contribute and make taxonomy more effective, as is the case for Coccinellidae. For precise identification of species of ladybird beetles, for example, it would be essential to generate genomic data from primary types from museum collections. To-date, only WGS, which has low coverage, has been used to study ladybird beetles from museums, and, as shown in our projects, the results are promising. With the new genomes available, it will soon be possible to integrate other sequencing methods such as hybrid enrichment or hybridization capture using RAD probes, which will undoubtedly optimize the quantity of data obtained selectively from specific regions of the genome. This will be highly useful in robustly determining the phylogenetic relations in this family.

7.7. References

Ali, O.A., O'Rourke, S.M., Amish, S.J., Meek, M.H., Luikart, G., Jeffres, C., Miller, M.R. (2016). RAD capture (Rapture): Flexible and efficient sequence-based genotyping. *Genetics*, 202, 389–400. doi: 10.1534/genetics.115.183665.

Andersen, J.C. and Mills, N.J. (2012). DNA extraction from museum specimens of parasitic Hymenoptera. *PLoS ONE*, 7, e45549. doi: 10.1371/journal.pone.0045549.

Ariño, A.H. (2010). Approaches to estimating the universe of natural history collections data. *Biodiversity Informatics*, 7, 81–92.

Benham, P.M. and Bowie, R.C.K. (2023). Natural history collections as a resource for conservation genomics: Understanding the past to preserve the future. *Journal of Heredity*, 114, 367–384. doi: 10.1093/jhered/esac066.

Berger, A., Petschenka, G., Degenkolb, T., Geisthardt, M., Vilcinskas, A. (2021). Insect collections as an untapped source of bioactive compounds – Fireflies (Coleoptera: Lampyridae) and cardiotonic steroids as a proof of concept. *Insects*, 12, 689. doi: 10.3390/insects12080689.

Blaimer, B.B., Lloyd, M.W., Guillory, W.X., Brady, S.G. (2016). Sequence capture and phylogenetic utility of genomic ultraconserved elements obtained from pinned insect specimens. *PLoS ONE*, 11, e0161531. doi: 10.1371/journal.pone.0161531.

Brasil, S.N.R., Kelemen, E.P., Rehan, S.M. (2023). Historic DNA uncovers genetic effects of climate change and landscape alteration in two wild bee species. *Conservation Genetics*, 24, 85–98. doi: 10.1007/s10592-022-01488-w.

Buenaventura, E., Lloyd, M.W., Perilla López, J.M., González, V.L., Thomas-Cabianca, A., Dikow, T. (2021). Protein-encoding ultraconserved elements provide a new phylogenomic perspective of Oestroidea flies (Diptera: Calyptratae). *Systematic Entomology*, 46, 5–27. doi: 10.1111/syen.12443.

Burrell, A.S., Disotell, T.R., Bergey, C.M. (2015). The use of museum specimens with high-throughput DNA sequencers. *Journal of Human Evolution*, 79, 35–44. doi: 10.1016/j.jhevol.2014.10.015.

Call, E., Mayer, C., Twort, V., Dietz, L., Wahlberg, N., Espeland, M. (2021). Museomics: Phylogenomics of the moth family Epicopeiidae (Lepidoptera) using target enrichment. *Insect Systematics and Diversity*, 5, 1–10. doi: 10.1093/isd/ixaa021.

Cambefort, Y. (2006). *Des coléoptères, des collections et des hommes*. Muséum national d'Histoire naturelle, Paris. doi: 10.4000/books.mnhn.2265.

Card, D.C., Shapiro, B., Giribet, G., Moritz, C., Edwards, S.V. (2021). Museum genomics. *Annual Review of Genetics*, 55, 633–659. doi: 10.1146/annurev-genet-071719-020506.

Che, L., Zhang, P., Deng, S., Escalona, H.E., Wang, X., Li, Y., Pang, H., Vandenberg, N., Ślipiński, A., Tomaszewska, W. (2021). New insights into the phylogeny and evolution of lady beetles (Coleoptera: Coccinellidae) by extensive sampling of genes and species. *Molecular Phylogenetics and Evolution*, 156, 107045. doi: 10.1016/j.ympev.2020.107045.

Cong, Q., Shen, J., Zhang, J., Li, W., Kinch, L.N., Calhoun, J.V., Warren, A.D., Grishin, N.V. (2021). Genomics reveals the origins of historical specimens. *Molecular Biology and Evolution*, 38, 2166–2176. doi: 10.1093/molbev/msab013.

Crowley, L.M. (2021). The genome sequence of the seven-spotted ladybird, *Coccinella septempunctata* Linnaeus, 1758. *Wellcome Open Research*, 6, 319. doi: 10.12688/wellcomeopenres.17346.1.

Crowley, L.M., Barclay, M., Roy, H.E., Brown, P.M.J. (2023). The genome sequence of the orange ladybird, *Halyzia sedecimguttata* (Linnaeus, 1758). *Wellcome Open Research*, 8, 186. doi: 10.12688/wellcomeopenres.19369.1.

De-Dios, T., Fontsere, C., Renom, P., Stiller, J., Llovera, L., Uliano-Silva, M., Sánchez-Gracia, A., Wright, C., Lizano, E., Caballero, B. et al. (2023). Whole-genomes from the extinct Xerces Blue butterfly can help identify declining insect species. *eLife*, 12, RP87928. doi: 10.7554/eLife.87928.1.

Escalona, H.E., Zwick, A., Li, H-S., Li, J., Wang, X., Pang, H., Hartley, D., Jermiin, L.S., Nedvěd, O., Misof, B. et al. (2017). Molecular phylogeny reveals food plasticity in the evolution of true ladybird beetles (Coleoptera: Coccinellidae: Coccinellini). *BMC Evolutionary Biology*, 17, 151. doi: 10.1186/s12862-017-1002-3.

Espeland, M., Irestedt, M., Johanson, K., Åkerlund, M., Bergh, J.-E., Källersjö, M. (2010). Dichlorvos exposure impedes extraction and amplification of DNA from insects in museum collections. *Frontiers in Zoology*, 7, 2. doi: 10.1186/1742-9994-7-2.

Gauthier, J., Pajkovic, M., Neuenschwander, S., Kaila, L., Schmid, S., Orlando, L., Alvarez, N. (2020). Museomics identifies genetic erosion in two butterfly species across the 20th century in Finland. *Molecular Ecology Resources*, 20, 1191–1205. doi: 10.1111/1755-0998.13167.

Gauthier, J., Borer, M., Toussaint, E.F.A., Bilat, J., Kippenberg, H., Alvarez, N. (2023). Museomics reveals evolutionary history of *Oreina* alpine leaf beetles (Coleoptera: Chrysomelidae). *Systematic Entomology*, 1–14. doi: 10.1111/syen.12601.

Gilbert, M.T.P., Moore, W., Melchior, L., Worobey, M. (2007). DNA extraction from dry museum beetles without conferring external morphological damage. *PLoS ONE*, 2, e272. doi: 10.1371/journal.pone.0000272.

Goate, Z. (2022). The genome sequence of the two-spot ladybird, *Adalia bipunctata* (Linnaeus, 1758). *Wellcome Open Research*, 7, 288. doi: 10.12688/wellcome openres.18610.1.

Goldstein, P.Z. and Desalle, R. (2003). Calibrating phylogenetic species formation in a threatened insect using DNA from historical specimens. *Molecular Ecology*, 12, 1993–1998. doi: 10.1046/j.1365-294X.2003.01860.x.

Goodman, A., Tolman, E., Uche-Dike, R., Abbott, J., Breinholt, J.W., Bybe, S., Frandsen, P.B., Gosnell, J.S., Guralnick, R., Kalkman, V.J. et al. (2023). Assessment of targeted enrichment locus capture across time and museums using odonate specimens. *Insect Systematics and Diversity*, 7, 1–9. doi: 10.1093/isd/ixad011.

Goodwin, S., McPherson, J.D., McCombie, W.R. (2016). Coming of age: Ten years of next-generation sequencing technologies. *Nature Reviews Genetics*, 17, 333–351. doi: 10.1038/nrg.2016.49.

Grewe, F., Kronforst, M.R., Pierce, N.E., Moreau, C.S. (2021). Museum genomics reveals the *Xerces* blue butterfly (*Glaucopsyche xerces*) was a distinct species driven to extinction. *Biology Letters*, 17, 20210123. doi: 10.1098/rsbl.2021.0123.

Harper, G.L., Maclean, N., Goulson, D. (2006). Analysis of museum specimens suggests extreme genetic drift in the Adonis blue butterfly (*Polyommatus bellargus*). *Biological Journal of the Linnean Society*, 88, 447–452. doi: 10.1111/j.1095-8312.2006.00632.x.

Hartley, C.J., Newcomb, R.D., Russell, R.J., Yong, C.G., Stevens, J.R., Yeates, D.K., La Salle, J., Oakeshott, J.G. (2006). Amplification of DNA from preserved specimens shows blowflies were preadapted for the rapid evolution of insecticide resistance. *Proceedings of the National Academy of Sciences*, 103, 8757–8762. doi: 10.1073/pnas.0509590103.

Heintzman, P.D., Elias, S.A., Moore, K., Paszkiewicz, K., Barnes, I. (2014). Characterizing DNA preservation in degraded specimens of Amara alpina (Carabidae: Coleoptera). *Molecular Ecology Resources*, 14, 606–615. doi: 10.1111/1755-0998.12205.

Huang, W., Xie, X., Liang, X., Wang, X., Chen, X. (2019). Effects of different pretreatments of DNA extraction from dried specimens of ladybird beetles (Coleoptera: Coccinellidae). *Insects*, 10, 91. doi: 10.3390/insects10040091.

Huang, W., Xie, X., Huo, L., Liang, X., Wang, X., Chen, X. (2020). An integrative DNA barcoding framework of ladybird beetles (Coleoptera: Coccinellidae). *Scientific Reports*, 10, 10063. doi: 10.1038/s41598-020-66874-1.

Hundsdoerfer, A.K. and Kitching, I.J. (2017). Historic DNA for taxonomy and conservation: A case-study of a century-old Hawaiian hawkmoth type (Lepidoptera: Sphingidae). *PLoS ONE*, 12, e0173255. doi: 10.1371/journal.pone.0173255.

Jin, M., Zwick, A., Ślipiński, A., Keyzer, R., Pang, H. (2020). Museomics reveals extensive cryptic diversity of Australian prionine longhorn beetles with implications for their classification and conservation. *Systematic Entomology*, 45, 745–770. doi: 10.1111/syen.12424.

Johnston, M.A., Smith, A.D., Kanda, K., Kamiński, M.J., Naverette, P., Sanchez, L.A., Aalbu, R.L., Miller, K.B., Wheeler, Q.D., Franz, N.M. (2022). Testing the taxonomy of Amphidorini Leconte (Coleoptera: Tenebrionidae): A molecular phylogeny leveraging museum sequencing. *Annales Zoologici*, 72, 49–68. doi: 10.3161/00034541ANZ2022.72.1.003.

Kanda, K., Pflug, J.M., Sproul, J.S., Dasenko, M.A., Maddison, D.R. (2015). Successful recovery of nuclear protein-coding genes from small insects in museums using Illumina sequencing. *PLoS ONE*, 10, e0143929. doi: 10.1371/journal.pone.0143929.

Lemmon, A.R., Emme, S.A., Lemmon, E.M. (2012). Anchored hybrid enrichment for massively high-throughput phylogenomics. *Systematic Biology*, 61, 727–744. doi: 10.1093/sysbio/sys049.

Li, W., Łączyński, P., Escalona, H.E., Eberle, J., Huo, L., Chen, X., Huang, W., Chen, B., Ahrens, D., Ślipiński, A. et al. (2020). Combined molecular and morphological data provide insights into the evolution and classification of Chilocorini ladybirds (Coleoptera: Coccinellidae). *Systematic Entomology*, 45, 447–463. doi: 10.1111/syen.12409.

Linard, B., Arribas, P., Andújar, C., Crampton-Platt, A., Vogler, A.P. (2016). Lessons from genome skimming of arthropod-preserving ethanol. *Molecular Ecology Resources*, 16, 1365–1377. doi: 10.1111/1755-0998.12539.

Linderholm, A. (2016). Ancient DNA: The next generation – chapter and verse. *Biological Journal of the Linnean Society*, 117, 150–160. doi: 10.1111/bij.12616.

Lopez, L., Turner, K.G., Bellis, E.S., Lasky, J.R. (2020). Genomics of natural history collections for understanding evolution in the wild. *Molecular Ecology Resources*, 20, 1153–1160. doi: 10.1111/1755-0998.13245.

Majerus, M.E.N. (2016). *A Natural History of Ladybird Beetles*. Cambridge University Press, Cambridge. doi: 10.1017/CBO9781316336960.

Matos-Maraví, P., Duarte Ritter, C., Barnes, C.J., Nielsen, M., Olsson, U., Wahlberg, N., Marquina, D., Sääksjärvi, I., Antonelli, A. (2019). Biodiversity seen through the perspective of insects: 10 simple rules on methodological choices and experimental design for genomic studies. *Peer Journal*, 7, e6727. doi: 10.7717/peerj.6727.

Mayer, C., Dietz, L., Call, E., Kukowka, S., Martin, S., Espeland, M. (2021). Adding leaves to the Lepidoptera tree: Capturing hundreds of nuclear genes from old museum specimens. *Systematic Entomology*, 46, 649–671. doi: 10.1111/syen.12481.

Meredith, L.A. (1996). Roles of natural history collections. *Annals of the Missouri Botanical Garden*, 83, 536–545. doi: 10.2307/1221588.

Mikheyev, A.S., Tin, M.M.Y., Arora, J., Seeley, T.D. (2015). Museum samples reveal rapid evolution by wild honey bees exposed to a novel parasite. *Nature Communications*, 6, 7991. doi: 10.1038/ncomms8991.

Mikheyev, A.S., Zwick, A., Magrath, M.J.L., Grau, M.L., Qiu, L., Su, Y.N., Yeates, D. (2017). Museum genomics confirms that the Lord Howe Island stick insect survived extinction. *Current Biology*, 27, 3157–3161.e4. doi: 10.1016/j.cub.2017.08.058.

Miller, S.E., Barrow, L.N., Ehlman, S.M., Goodheart, J.A., Greiman, S.E., Lutz, H.L., Misiewicz, T.M., Smith, S.M., Tan, M., Thawley, C.J. et al. (2020). Building natural history collections for the twenty-first century and beyond. *BioScience*, 70(8), 674–687. doi: 10.1093/biosci/biaa069.

Moris, V.C., Christmann, K., Wirtgen, A., Belokobylskij, S.A., Berg, A., Liebig, W.H., Soon, V., Baur, H., Schmitt, T., Niehuis, O. (2021). Cuticular hydrocarbons on old museum specimens of the spiny mason wasp, *Odynerus spinipes* (Hymenoptera: Vespidae: Eumeninae), shed light on the distribution and on regional frequencies of distinct chemotypes. *Chemoecology*, 31, 311–322. doi: 10.1007/s00049-021-00350-8.

Mullin, V.E., Stephen, W., Arce, A.N., Nash, W., Raine, C., Notton, D.G., Whiffin, A., Blagderov, V., Gharb, I.K., Hogan, J. et al. (2023). First large-scale quantification study of DNA preservation in insects from natural history collections using genome-wide sequencing. *Methods in Ecology and Evolution*, 14, 360–371. doi: 10.1111/2041-210X.13945.

Nagy, Z.T. (2010). A hands-on overview of tissue preservation methods for molecular genetic analyses. *Organisms Diversity & Evolution*, 10, 91–105. doi: 10.1007/s13127-010-0012-4.

Nakahama, N., Isagi, Y., Ito, M. (2019). Methods for retaining well-preserved DNA with dried specimens of insects. *European Journal of Entomology*, 116, 486–491. doi: 10.14411/EJE.2019.050.

Nakazato, T. and Jinbo, U. (2022). Cross-sectional use of barcode of life data system and GenBank as DNA barcoding databases for the advancement of museomics. *Frontiers in Ecology and Evolution*, 10, 966605. doi: 10.3389/fevo.2022.966605.

Nattier, R. (2018). Biodiversity in natural history collections: A source of data for the study of evolution. In *Biodiversity and Evolution*, Grandcolas, P. and Maurel, M.-C. (eds). ISTE Ltd, London, and John Wiley & Sons, New York.

Nattier, R. and Salazar, K. (2019). Next-generation sequencing yields mitochondrial genome of *Coccidophilus cariba* Gordon (Coleoptera: Coccinellidae) from museum specimen. *Mitochondrial DNA Part B*, 4, 3780–3781. doi: 10.1080/23802359.2019.1679684.

Nattier, R., Michel-Salzat, A., Almeida, L.M., Chifflet-Belle, P., Magro, A., Salazar, K., Kergoat, G.J. (2021). Phylogeny and divergence dating of the ladybird beetle tribe Coccinellini Latreille (Coleoptera: Coccinellidae: Coccinellinae). *Systematic Entomology*, 46, 632–648. doi: 10.1111/syen.12480.

NCBI (1988). National Library of Medicine (US), National Center for Biotechnology Information, Bethesda, MD [Online]. Available at: https://www.ncbi.nlm.nih.gov/ [Accessed 7 June 2023].

Ney, G., Frederick, K., Schul, J. (2018). A post-pleistocene calibrated mutation rate from insect museum specimens. *PLoS Currents*, 10. doi: 10.1371/currents.tol.aba557de56be881793261f7e1565cf35.

Nunes, R., Storer, C., Doleck, T., Kawahara, A.Y., Pierce, N.E., Lohman, D.J. (2022). Predictors of sequence capture in a large-scale anchored phylogenomics project. *Frontiers in Ecology and Evolution*, 10, 943361. doi: 10.3389/fevo.2022.943361.

Pääbo, S., Poinar, H., Serre, D., Jaenicke-Després, V., Hebler, J., Rohland, N., Kuch, M., Krause, J., Vigilant, L., Hofreiter, M. (2004). Genetic analyses from ancient DNA. *Annual Review of Genetics*, 38, 645–679. doi: 10.1146/annurev.genet.37.110801.143214.

Parejo, M., Wragg, D., Henriques, D., Charrière, J.-D., Estonba, A. (2020). Digging into the genomic past of Swiss honey bees by whole-genome sequencing museum specimens. *Genome Biology and Evolution*, 12, 2535–2551. doi: 10.1093/gbe/evaa188.

Patzold, F., Zilli, A., Hundsdoerfer, A.K. (2020). Advantages of an easy-to-use DNA extraction method for minimal-destructive analysis of collection specimens. *PLoS ONE*, 15, e0235222. doi: 10.1371/journal.pone.0235222.

Pitteloud, C., Arrigo, N., Suchan, T., Mastretta-Yanes, A., Vila, R., Dincă, V., Hernández-Roldán, J., Brockmann, E., Chittaro, Y., Kleckova, I. et al. (2017). Climatic niche evolution is faster in sympatric than allopatric lineages of the butterfly genus *Pyrgus*. *Proceedings of the Royal Society B: Biological Sciences*, 284, 20170208. doi: 10.1098/rspb.2017.0208.

Price, B.W., Henry, C.S., Hall, A.C., Mochizuki, A., Duelli, P., Brooks, S.J. (2015). Singing from the grave: DNA from a 180 year old type specimen confirms the identity of *Chrysoperla carnea* (Stephens). *PLoS ONE*, 10, e0121127. doi: 10.1371/journal.pone.0121127.

Ratnasingham, S. and Hebert, P.D.N. (2013). A DNA-based registry for all animal species: The Barcode Index Number (BIN) system. *PLoS ONE*, 8, e66213. doi: 10.1371/journal.pone.0066213.

Raxworthy, C.J. and Smith, B.T. (2021). Mining museums for historical DNA: Advances and challenges in museomics. *Trends in Ecology & Evolution*, 36, 1049–1060. doi: 10.1016/j.tree.2021.07.009.

Rizzi, E., Lari, M., Gigli, E., De Bellis, G., Caramelli, D. (2012). Ancient DNA studies: New perspectives on old samples. *Genetics Selection Evolution*, 44, 21. doi: 10.1186/1297-9686-44-21.

Ryan, S.F., Deines, J.M., Scriber, J.M., Pfrender, M.E., Jones, S.E., Emrich, S.J., Hellmann, J.J. (2018). Climate-mediated hybrid zone movement revealed with genomics, museum collection, and simulation modeling. *Proceedings of the National Academy of Sciences*, 115, E2284–E2291. doi: 10.1073/pnas.1714950115.

Salazar, K. and Nattier, R. (2020). New light on historical specimens reveals a new species of ladybird (Coleoptera: Coccinellidae): Morphological, museomic, and phylogenetic analyses. *Insects*, 11, 766. doi: 10.3390/insects11110766.

Schmid, S., Genevest, R., Gobet, E., Suchan, T., Sperisen, C., Tinner, W., Alvarez, N. (2017). HyRAD-X, a versatile method combining exome capture and RAD sequencing to extract genomic information from ancient DNA. *Methods in Ecology and Evolution*, 8, 1374–1388. doi: 10.1111/2041-210X.12785.

Shaffer, H.B., Fisher, R.N., Davidson, C. (1998). The role of natural history collections in documenting species declines. *Trends in Ecology and Evolution*, 13, 27–30.

Shpak, M., Ghanavi, H.R., Lange, J.D., Pool, J.E., Stensmyr, M.C. (2023). Genomes from historical *Drosophila melanogaster* specimens illuminate adaptive and demographic changes across more than 200 years of evolution. *PLoS Biology*, 21, e3002333. doi: 10.1371/journal.pbio.3002333.

Ślipiński, A. (2007). *Australian Ladybird Beetles (Coleoptera: Coccinellidae): Their Biology and Classification*. ABRS/CSIRO Publishing, Canberra/Melbourne.

Sproul, J.S. and Maddison, D.R. (2017). Sequencing historical specimens: Successful preparation of small specimens with low amounts of degraded DNA. *Molecular Ecology Resources*, 17, 1183–1201. doi: 10.1111/1755-0998.12660.

Sproul, J.S. and Maddison, D.R. (2018). Cryptic species in the mountaintops: Species delimitation and taxonomy of the *Bembidion breve* species group (Coleoptera: Carabidae) aided by genomic architecture of a century-old type specimen. *Zoological Journal of the Linnean Society*, 183, 556–583. doi: 10.1093/zoolinnean/zlx076.

St Laurent, R.A., Hamilton, C.A., Kawahara, A.Y. (2018). Museum specimens provide phylogenomic data to resolve relationships of sack-bearer moths (Lepidoptera, Mimallonoidea, Mimallonidae). *Systematic Entomology*, 43, 729–761. doi: 10.1111/syen.12301.

Straub, S.C.K., Parks, M., Weitemier, K., Fishbein, M., Cronn, R.C., Liston, A. (2012). Navigating the tip of the genomic iceberg: Next-generation sequencing for plant systematics. *American Journal of Botany*, 99, 349–364. doi: 10.3732/ajb.1100335.

Suchan, T., Pitteloud, C., Gerasimova, N.S., Kostikova, A., Schmid, S., Arrigo, N., Pajkovic, M., Ronikier, M., Alvarez, N. (2016). Hybridization capture using RAD probes (hyRAD), a new tool for performing genomic analyses on collection specimens. *PLoS ONE*, 11, e0151651. doi: 10.1371/journal.pone.0151651.

Tay, W.T., Elfekih, S., Polaszek, A., Court, L.N., Evans, G.A., Gordon, K.H.J., De Barro, P.J. (2017). Novel molecular approach to define pest species status and tritrophic interactions from historical *Bemisia* specimens. *Scientific Reports*, 7, 429. doi: 10.1038/s41598-017-00528-7.

Thomsen, P.F., Elias, S., Gilbert, M.T.P., Haile, J., Munch, K., Kuzmina, S., Froese, D.G., Sher, A., Holdaway, R.N., Willerslev, E. (2009). Non-destructive sampling of ancient insect DNA. *PLoS ONE*, 4, e5048. doi: 10.1371/journal.pone.0005048.

Tin, M.M.-Y.Y., Economo, E.P., Mikheyev, A.S. (2014). Sequencing degraded DNA from non-destructively sampled museum specimens for RAD-tagging and low-coverage shotgun phylogenetics. *PLoS ONE*, 9, e96793. doi: 10.1371/journal.pone.0096793.

Tomaszewska, W., Escalona, H.E., Hartley, D., Li, J., Wang, X., Li, H., Pang, H., Ślipiński, A., Zwick, A. (2021). Phylogeny of true ladybird beetles (Coccinellidae: Coccinellini) reveals pervasive convergent evolution and a rapid Cenozoic radiation. *Systematic Entomology*, 46, 611–631. doi: 10.1111/syen.12479.

Toussaint, E.F.A., Gauthier, J., Bilat, J., Gillett, C.P.D.T., Gough, H.M., Lundkvist, H., Blanc, M., Muñoz-Ramírez, C.P., Alvarez, N. (2021). HyRAD-X exome capture museomics unravels giant ground beetle evolution. *Genome Biology and Evolution*, 13, evab112. doi: 10.1093/gbe/evab112.

Townson, H., Harbach, R.E., Callan, T.A. (1999). DNA identification of museum specimens of the *Anopheles gambiae* complex: An evaluation of PCR as a tool for resolving the formal taxonomy of sibling species complexes. *Systematic Entomology*, 24, 95–100. doi: 10.1046/j.1365-3113.1999.00084.x.

Twort, V.G., Minet, J., Wheat, C.W., Wahlberg, N. (2021). Museomics of a rare taxon: Placing whalleyanidae in the lepidoptera tree of life. *Systematic Entomology*, 46, 926–937. doi: 10.1111/syen.12503.

Van Dam, M.H., Lam, A.W., Sagata, K., Gewa, B., Laufa, R., Balke, M., Faircloth, B.C., Riedel, A. (2018). Ultraconserved elements (UCEs) resolve the phylogeny of Australasian smurf-weevils. *PLoS ONE*, 13, e0205049. doi: 10.1371/journal.pone.0205049.

Wandeler, P., Hoeck, P.E.A., Keller, L.F. (2007). Back to the future: Museum specimens in population genetics. *Trends in Ecology & Evolution*, 22, 634–42. doi: 10.1016/j.tree.2007.08.017.

Yamashita, Y., Ogura-Tsujita, Y., Nagata, N., Kurosawa, T., Yukawa, T. (2022). Molecular identification of seed-feeding flies dissected from herbarium specimens clarifies the 100-year history of parasitism by *Japanagromyza tokunagai* in Japan. *Ecological Research*, 37, 240–256. doi: 10.1111/1440-1703.12283.

Yeates, D.K., Zwick, A., Mikheyev, A.S. (2016). Museums are biobanks: Unlocking the genetic potential of the three billion specimens in the world's biological collections. *Current Opinion in Insect Science*, 18, 83–88. doi: 10.1016/J.COIS.2016.09.009.

Zhang, J., Cong, Q., Shen, J., Brockmann, E., Grishin, N.V. (2019). Genomes reveal drastic and recurrent phenotypic divergence in firetip skipper butterflies (Hesperiidae: Pyrrhopyginae). *Proceedings of the Royal Society B: Biological Sciences*, 286, 20190609. doi: 10.1098/rspb.2019.0609.

Zhou, Y.L., Nicholls, J.A., Liu, Z.H., Hartley, D., Szito, A., Slipiński, A., Zwick, A. (2022). Molecular phylogeny of Dermestidae (Coleoptera) reveals the polyphyletic nature of *Trogoderma* Latreille and the taxonomic placement of the Khapra Beetle *Trogoderma granarium* Everts. *Insect Systematics and Diversity*, 6, 1–24. doi: 10.1093/isd/ixac026.

Zhu, W., Chi, S., Wang, Y., Li, H., Wang, Z., Gu, S., Sun, T., Xiang, H., You, P., Ren, Y. (2023). A chromosome-level genome assembly of the *Henosepilachna vigintioctomaculata* provides insights into the evolution of ladybird beetles. *DNA Research: An International Journal for Rapid Publication of Reports on Genes and Genomes*, 30, 1–11. doi: 10.1093/dnares/dsad001.

8

New Tools and New Discoveries in Paleo-entomology: Looking to Future Challenges

8.1. Insects as essential players in past and present ecosystems

The most diversified known class of animals is Hexapoda ("insects"), with over 1,100,000 modern species described, representing a huge 75% of the entirety of animal diversity (Stork 2018; Sollai and Solari 2022). They occupy practically all ecological niches in terrestrial and aquatic ecosystems.

Few taxa are restricted to coastal or pelagic marine environments (for example, Heteroptera Gerridae Halobatinae). Insects make use of all available resources: they are carnivores; hematophages; herbivores, including pollinivores and nectarivores; recyclers of plant remnants; necrophages; coprophages; parasites; and parasitoids.

Insects have social organizations of varying complexity. Certain species take care of their young (e.g. Coleoptera Passalidae, Hemiptera Aphididae); others have immense and vastly complex societies (e.g. in Hymenoptera, Isoptera). Their unparalleled diversity, ecological and morphofunctional disparity and their presence within all trophic groups mean Hexapoda are essential ecosystemic actors. Thus, it is crucial to study insects in order to gain an understanding of how the modern biosphere was established, and of its workings.

Chapter written by André NEL, Patricia NEL, Jean-Paul KUNDURA, Romain GARROUSTE, Valérie NGÔ-MULLER, Mathieu BODERAU, Thomas SCHUBNEL and Corentin JOUAULT.

Time-calibrated phylogenetic analyses suggest that a divergence between Ectognatha (= Insecta) and other Hexapoda occurred during the Ordovician-Silurian transition (for example, see Misof et al. 2014; Tong et al. 2015; Montagna et al. 2019). Unfortunately, there is some controversy as to their most ancient representatives, sometimes voiced with utmost zeal (e.g. by Garrouste et al. 2012; or Hörnschemeyer et al. 2013). Hexapoda are among the most ancient animals to have "colonized" the land. They may even pre-date land-dwelling vertebrates, although Arachnida and Myriapoda, which have been found to exist since the Ordovician and Silurian, came before the most ancient insects. Whilst Devonian and lower Carboniferous insects are yet unknown, there was considerable entomic diversity from the beginning of the upper Carboniferous. With over 28,200 fossil species described (between the Carboniferous and the Holocene), they represent 17% of animal paleodiversity (calculations using the Paleobiology Database[1]) and are the animal group with the largest known number of fossil species.

By comparing their diversity in exceptionally well-preserved deposits dating from the upper Carboniferous, we see that they were extremely varied, occupying key places in the trophic chains even during that period (Grimaldi and Engel 2005).

Contrary to the ideas which were widespread until the 2000s, fossilized insects are abundant, often very well preserved, and provide astonishingly complete morphological information (see Figure 8.1), compared to fossilized vertebrates, which are often known for the preservation of only their teeth; almost never skin and soft parts.

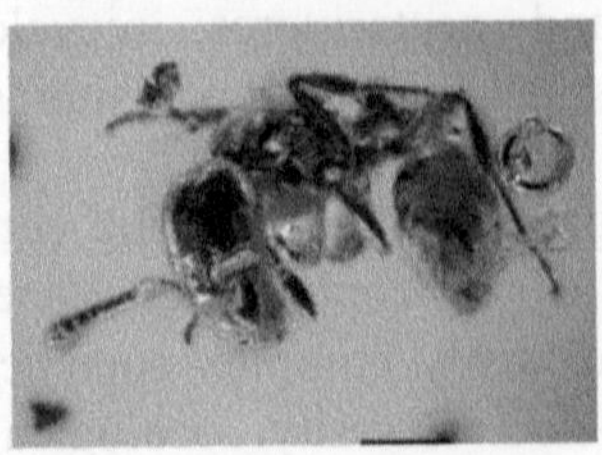

Figure 8.1. *An ant of the genus Fallomyrma sp. (Baltic amber, Priabonian) (collected by CJ). Scale = 0.5 mm. For a color version of this figure, see www.iste.co.uk/nicolas/biodiversity.zip*

1 Available at: www.paleobiodb.org/#/ [Accessed 26 March 2024].

Insects are generally found in lacustrine sediments or trapped in fossilized resins. More rarely, we see fossils in gypsum, selenite, sedimentary quartz and phosphorite. Frequently, the whole specimen is preserved, especially in resins, phosphorites and nodules. Due to chitin's resistance to putrefaction and the small size of insects, a "standard" source is likely to provide tens, or even hundreds, of thousands of specimens. This means that, for a much smaller investment in terms of effort, time and funding than that needed for the discovery of just one dinosaur, a paleo-entomologist may obtain thousands of fossil data points for study.

8.2. Discovering the past to understand the present (and perhaps predict the future)

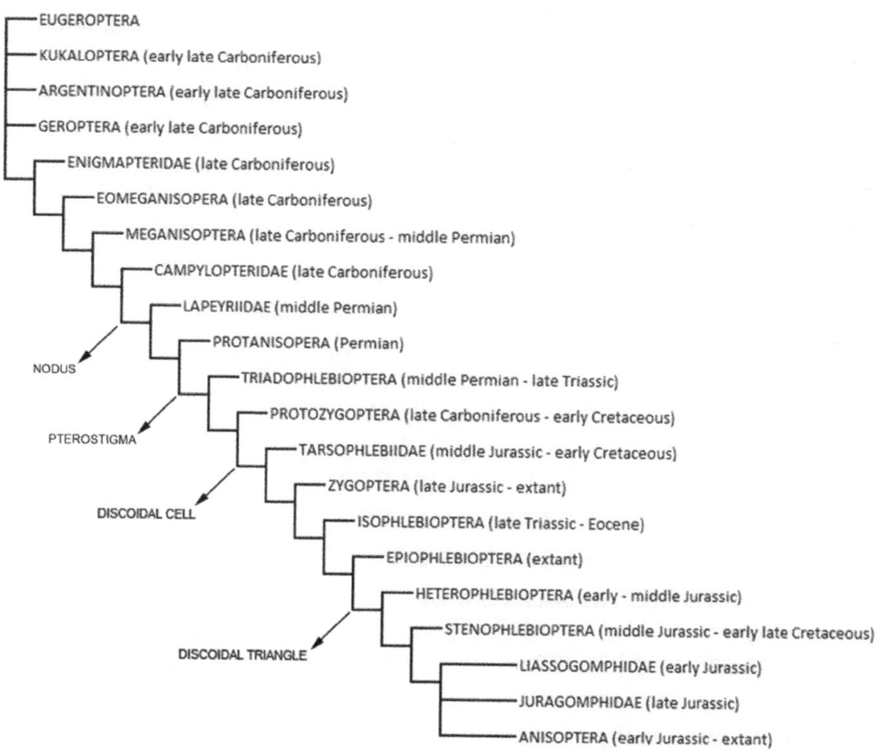

Figure 8.2. *Phylogenetic hypothesis for Odonatoptera (superorder containing modern Odonata)*

This aphorism is particularly true in relation to studies of insects, because this extremely ancient group plays a key role in ecosystems. Phylogenetic studies, including fossilized entomofauna have revealed that, by studying only taxa alive today, we are able to understand only a tiny part of the evolutionary history of the clades (for example, see Jouault et al. 2022a). Odonatoptera are an obvious example of this (see Figure 8.2) (Nel and Piney 2023).

8.2.1. *A science which, until recently, had been largely overlooked*

It was not until the 21st Century that paleo-entomology began to be seen as "useful" rather than as an exotic discipline exercised by a few eccentrics. This is all the more surprising, given that Willi Hennig, the father of modern phylogenetics, included fossils in his analyses, and produced a seminal work on the evolutionary history of insects (Hennig 1981), after the taxonomic works done by Scudder, Brongniart, Handlirsch and Rodendorf. In the second half of the 20th Century, paleo-entomology was studied mainly in a single institute in Moscow.

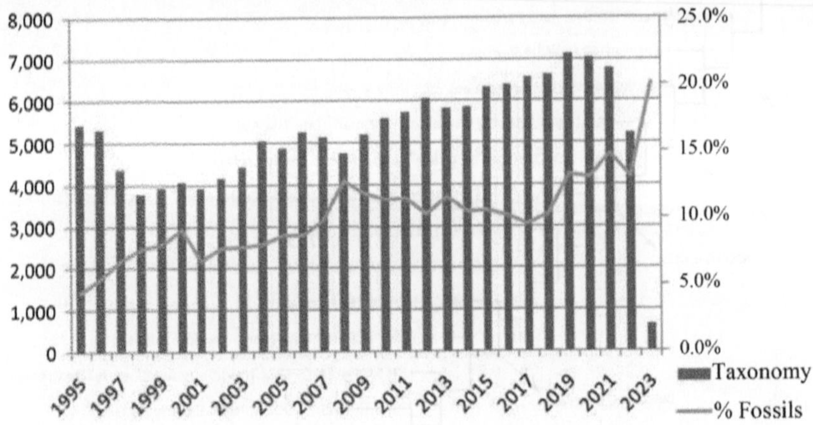

Figure 8.3. *Number of publications in taxonomy of insects and the proportion in relation to fossils (analysis by J.-P.K, based on the Zoological Record). For a color version of this figure, see www.iste.co.uk/nicolas/biodiversity.zip*

Things changed after the turn of the millennium, as attested by the spectacular rise in the relative number of publications devoted to fossilized insects (which, since 2008, represented between 10 and 13% of publications on the systematics of insects) and the number of species described (see Figure 8.3).

The effects of "fashion" partly account for this phenomenon (increased interest in fauna from Cretaceous resins – particularly those from Myanmar) (see Figure 8.4); partly, but not entirely. Research involving studies of fossilized insects is not limited to α taxonomy, but such specimens have been increasingly included in phylogenetic analyses, aiming to better understand the evolutionary history of various groups, or even of the entire clade (for example, see Ronquist et al. 2012; Vea and Grimaldi 2016; Jouault et al. 2021).

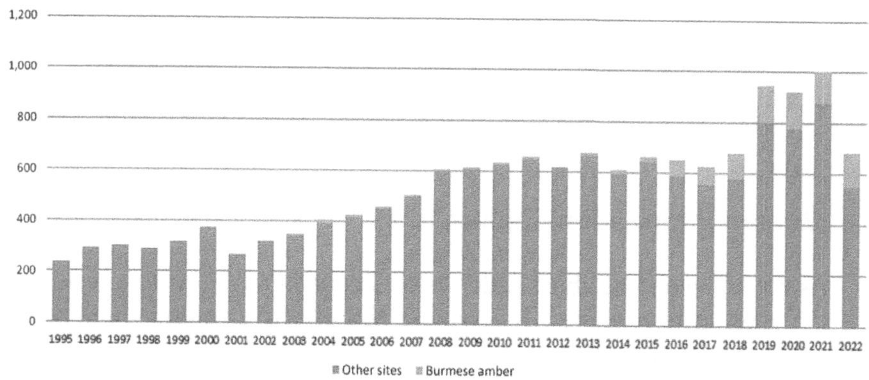

Figure 8.4. *Number of publications on fossilized insects per year (analysis by J.-P.K.; based on the Zoological Record). For a color version of this figure, see www.iste.co.uk/nicolas/biodiversity.zip*

8.2.2. *Spectacular recent progress in data analysis*

Research, on both modern and fossilized insects with homologous primary characteristics, is crucial for any phylogenetic analysis. For example, the wings (essential structures) are best preserved in fossilized insects. By comparing them with the wings of modern insects, using 3D imaging with X-ray tomography, researchers have been able to put forward a new model of their nervation (Desutter-Grandcolas et al. 2017; Jacquelin et al. 2018; Schubnel et al. 2020).

To begin with, fossils served as "calibration points" to date the occurrences of clades, but were not integrated as terminal taxa into time-calibrated phylogenetic analyses. It was only with the progress of Bayesian phylogenetic analyses and methodological improvements that dating analyses based on *total evidence* began to flourish. In such studies, fossils came to occupy their rightful place as organisms worthy of study in their own right, just like their modern-day "descendants". Models of the speciation and extinction processes can be used to put forward testable scenarios relating to the evolutionary changes that affected clades, to estimate the impacts of crises and compare them to those of more gradual changes in ecosystems (examples can be found in Jouault et al. 2022b, 2022c), thus avoiding limiting the reflection to unverifiable "scenarios" that are based on the researcher's "auctoritas". Other approaches that deal more with paleo-ecology, with testable results, are now available, based on rankings of the geological periods by the presence or absence of insects exhibiting particular morphological structures (the WAPUM method) (Nel et al. 2018).

8.3. Modern information-capture tools

It is not enough to be able to add fossils into integrative evolution studies; we still need to have materials to study, and effective information-capture tools. Too often, neontologists do not integrate fossils into their analyses, based on the argument – not always true – that there will be no fossils in the group in question, or that it is difficult to compare them to their modern "representatives".

8.3.1. *Bibliography: "I had a dream"*

Bibliographic tools are indispensable, and must relate to the fossil record, but also to modern taxa. In part, the Internet has helped break down the barrier of difficulty in accessing literature. The free Paleobiology Database[2], launched in 2005, is particularly useful and complete with an accreditation from the Web of Science[3]; this latter requires paid membership, and is not accessible to all researchers. However, there is still a great deal of skill involved in bibliography, given the copyright problems relating to various

2 See: www.fossilworks.org.

3 See: www.webofscience.com.

publishers. What is needed is the creation of a true database containing PDF copies of all entomological literature, freely available to all.

8.3.2. *The field*

The greatly increased rate of discoveries, since the 1990s, of new and extremely rich sites – particularly for fossils from the Cretaceous – has changed the game in terms of the supposed rarity of fossils. In Handlirsch's 1430-page treatise on fossilized insects (1908), only 13 pages were devoted to the Cretaceous. The sites of Liaoning (China), Crato (Brazil), amber deposits in Lebanon, France, Spain and – in particular – Myanmar, have revealed the extraordinary diversity and complexity of entomofauna from this key period in the establishment of modern terrestrial ecosystems. The discovery of new essential sites also related to other periods. Thus, field research carried out by a team of amateurs in the upper Carboniferous in Avion (France) completely changed our view of fossilized insects from this key period: there were entomofauna of very small insects, previously unknown, including the oldest known representatives of the clades currently most diversified: Hemiptera and Holometabola (Nel et al. 2013). These discoveries, since integrated into phylogenetic analyses, helped identify the main clades of insects as being considerably older than previously thought. Alas, this salvage of a fossil deposit which was doomed to destruction by the development of a slagheap, cannot conceal all the fossil sites which are lost to us through lack of proper management of France's paleontological heritage, similar to the management system in place for archeological heritage. This phenomenon is not a recent one: there have been no significant collections from a number of sites potentially very rich in Cretaceous amber, documented in the 19th Century; hence, this material is either lost or destroyed. What might have been excusable in the 19th century is less so now: local authorities or quarry operators prohibit access and salvaging of fossils from exceptional sites. How many old quarries have been "restored" without consulting paleontologists?

8.3.3. *Fossil preparation*

Note that the use of hot and/or cold exposure can help separate sedimentary layers and discover more fossils. The use of high-performance epoxy resins means that amber fragments can be included, though they are often fragile, before being polished. The first resins of this type had the terrible defect of heating the amber, causing it to shatter. In this way, a number of specimens were lost. However, progress has been made since then. The preparation process devised by Dany Azar (Azar et al. 1999), for extremely small and/or fragile pieces, is also of note.

8.3.4. *Information capture from fossils*

Considerable progress has been made recently in information capture from fossils, using digital imaging. Given their small size, fossilized insects are difficult to study, although extremely fine crucial structures may have been preserved (see Figure 8.5).

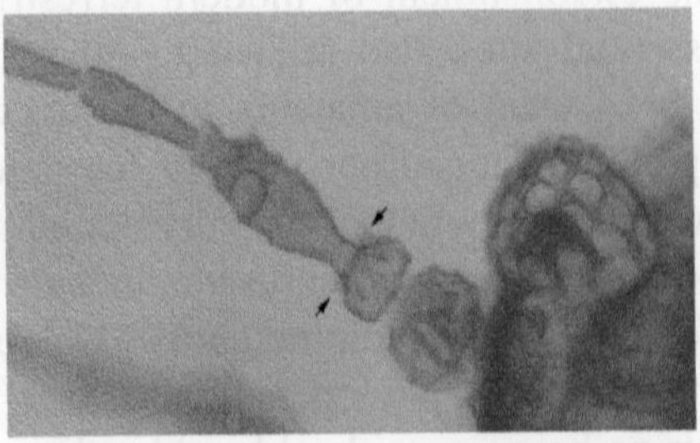

Figure 8.5. *Close-up view of the antenna of Thysanoptera Uzelothrips eocenicus, studied by Nel et al. 2013 (lower Eocene amber from Oise, France), showing the antennal sensilla. For a color version of this figure, see www.iste.co.uk/nicolas/biodiversity.zip*

In addition to the progress made in "conventional" optics (Reflectance Transforming Imaging – see Chen et al. 2021), new and highly innovative approaches have emerged since the turn of the millennium. One particular example is X-ray tomography. The earliest tests of this method were not able to detect the details of the wing venation of an insect in amber, but this barrier was overcome by collaboration between the MNHN and the ESRF, Grenoble (Tafforeau et al. 2006). Since then, the use of micro-CT scans has developed, and become (a little) more widely used. This technique allows us to see details on a scale of less than a micrometer, allowing for digital reconstruction of structures such as the male genitalia of a fossilized cricket or the anatomy of the head of a fossilized insect, in amber or mummified by compression in sediments (see Figure 8.6).

Figure 8.6. *Photograph and 3D reconstruction from microtomography of a Blattodea in lower Eocene amber from Oise (France) (currently in the process of description). For a color version of this figure, see www.iste.co.uk/nicolas/biodiversity.zip*

Observation under ultra-violet light, which is a technique known for yielding information about today's insects, also developed, in particular for insects trapped in fossilized resins (UV fluorescence, which offers a clearer view of the fine structures) (for example, see Fu et al. 2020). Very recently, we have been able to show UV fluorescence induced in fossilized insects compressed in sediments – something which is known to happen in the case of crustaceans (Barlow et al. 2021). In certain cases, structures which are very difficult to see under normal light become spectacularly clear under UV light (Nel et al. 2023a, 2023b). Environmental SEM (in a low vacuum) also allows access to morphological details without damaging the specimens.

8.4. More "exotic" approaches

8.4.1. *Fossil trophic interactions*

A new approach, developed by Conrad Labandeira and since implemented on numerous sites (Labandeira et al. 2007; Wappler et al. 2009), is the study of interactions between plants and animals. It allows us to better understand the changes which may have occurred in the ecosystems – especially around the times of the known major biodiversity crises. As the fossil record of plants may be more "continuous" than that of insects, by observing the traces of activity, we can analyze changes in the trophic chains before and after these crises. However, it is still limited by the difficulty in determining which taxa fed on plants. SEM study of pollen remnants in and on insects enables us to estimate the date of the interactions between the plants and their pollinators (for example, see Khramov et al. 2023).

8.4.2. *DNA and organic chemistry*

Old attempts to collect DNA from fossilized insects often met with failure. These avenues of research have been almost completely abandoned, though immense progress has been made in the field of ancient DNA. Very recently, Peris et al. (2020) showed that traces of DNA could be present in modern resins. It would certainly be worth trying new research on very well-preserved mummified insects in Miocene diatomites. While it seems that, for the time being, we must give up the hope of accessing the genomes of fossilized insects, there are fascinating prospects yet to be explored in the field of fossilized cuticle macromolecules, which could be used to identify incomplete fossils. It would be useful for Silurian or Devonian sites which have produced dislodged microfragments of cuticle. However, this approach would require active interactions between (paleo-)entomologists and biochemists. To-date, it has been applied only to plants, or in one case, to a scorpion (Cody et al. 2011; Braga et al. 2016; D'Angelo 2019). There is still a very great deal left to do on fossilized terrestrial arthropods. Recently, a synchrotron method (XRS) was used to produce a 3D chemical analysis of an insect preserved in amber (Georgiou et al. 2019), revealing a well-preserved chemical signature typical of chitinous polysaccharides, opening up an unexpected avenue in paleo-entomology (taphonomy, internal characteristics, paleobiology, etc.).

8.5. Conclusion

As a result of this progress in analysis and information-capture tools, the quantity and quality of the morphological characteristics available for the integration of fossils as terminal taxa in phylogenetic analyses have greatly increased, meaning that the results of these analyses are increasingly reliable. Consequently, there is less and less of a boundary between fossilized and modern taxa.

A new, realistic vision is developing, as to what a fossil is, as an organism that was once alive, and thus needing to be systematically integrated into studies on living organisms, thanks to recent developments of new analytical tools in phylogenetics. Thus, the study of evolution has finally (almost) managed to break out of the historical rut in which it was trapped by creationists of the 19th century (and sometimes even the 20th or 21st centuries), by separating research on fossils from research on modern-day specimens. However, in certain research organizations and museums, there is still a de facto separation between fossils and modern specimens – in particular, in the management of collections. One hopes that this state of affairs will not persist for too long.

We can no longer envisage the entomologists of the future as hyperspecialists in a particular group, working in ignorance of the existence of fossil taxa from that group. They need to be able to capture bibliographic data that are as exhaustive as possible, collect, prepare and study both modern and fossilized taxa, use 3D reconstructions, improve knowledge of modern and fossil insect anatomy, be able to work with phylogenetic tools rather than content themselves with producing phylogenies isolated from any comparison between the two information sources (which, in fact, constitute only one source: a comparative study of organisms). They can no longer study Paleozoic fossils as though they were "extraterrestrial" organisms with no connection to modern fauna, and with no concern for the monophyly of the groups that they propose.

This vision must ultimately be realized by the fusion or integration of purely paleontological research labs into institutes which encourage collaboration between researchers from multiple disciplines.

8.6. References

Azar, D., Nel, A., Solignac, M., Paicheler, J.-C., Bouchet, F. (1999). New genera and species of psychodoid flies from the Lower Cretaceous amber of Lebanon. *Palaeontology*, 42, 1101–1136.

Barlow, L.A., Pittman, M., Butcher, A., Martill, D.M., Kaye, T.G. (2021). Laser-stimulated fluorescence reveals unseen details in fossils from the Upper Jurassic Solnhofen Limestones. *Royal Society Open Sciences*, 8(211601), 1–13.

Braga, M.V., Mendonça, P.M., Barbosa, R.R.. Blomquist, G.J., Novo, S.P.C., Juliana da Matta Dutra, F., de Souza, S.M., de Carvalho Queiroz, M.M. (2016). Identification of *Megaselia scalaris* (Loew 1866) (Diptera: Phoridae) in mummified human body from Itacambira (MG), Brazil, using scanning electron microscopy and cuticular hydrocarbons. *Journal of Natural History*, 50(21/22), 1381–1388.

Chen, L., Blanke, A., Gu, J.-J., Yang, Q., Ren, D., Béthoux, O. (2021). Ovipositor and mouthparts in a new fossil insect support a novel ecological role for early orthopterans in Pennsylvanian forests. *eLife*, 10(e71006), 1–9.

Cody, G.D., Gupta1, N.S., Briggs, D.E.G., Kilcoyne, A.L.D., Summons, R.E., Kenig, F., Plotnick, R.E., Scott, A.C. (2011). Molecular signature of chitin-protein complex in Paleozoic arthropods. *Geology*, 39, 255–258.

D'Angelo, J.A. (2019). Molecular structure of the cuticles of *Dicroidium* and *Johnstonia* (Corystospermaceae, Triassic, Argentina). Ecophysiological adaptations of two chemically indistinguishable, morphology-based taxa. *Review of Palaeobotany and Palynology*, 268, 109–124.

Desutter-Grandcolas, L., Jacquelin, L., Hugel, S., Boistel, R., Garrouste, R., Henrotay, M., Warren, B.H., Chintauan-Marquier, I.C., Nel, P., Grandcolas, P. et al. (2017). 3-D imaging reveals four extraordinary cases of convergent evolution of acoustic communication in crickets and allies (Insecta). *Scientific Reports*, 7(1/7099), 18.

Fu, Y., Azar, D., Huang, D. (2020). A new species of the extinct family Minlagerrontidae (Insecta: Hemiptera: Cicadomorpha) from mid-Cretaceous Burmese amber. *Cretaceous Research*, 107(104270), 1–7.

Garrouste, R., Clément, G., Nel, P., Engel, M.S., Grandcolas, P., D'Haese, C., Lagebro, L., Denayer, J., Gueriau, P., Lafaite, P. et al. (2012). A complete insect from the Late Devonian period. *Nature*, 488, 82–85.

Georgiou, R., Gueriau, P., Sahle, C.J., Bernard, S., Mirone, A., Garrouste, R., Bergmann, U., Rueff, J.-P., Bertrand, L. (2019). Carbon speciation in organic fossils using 2D to 3D X-ray Raman multispectral imaging. *Science Advances*, 5(eaaw5019), 1–10.

Grimaldi, D.A. and Engel, M.S. (2005). *Evolution of the Insects*. Cambridge University Press, Cambridge.

Handlirsch, A. (1908). *Die fossilen Insekten und die Phylogenie der rezenten Formen. Ein Handbuch für Paläontologen und Zoologen*. Edition Leipzig, Leipzig.

Hennig, W. (1981). *Insect Phylogeny*. John Wiley & Sons, New York.

Hörnschemeyer, T., Haug, J., Bethoux, O., Beutel, R.G., Charbonnier, S., Hegna, T.A., Koch, M., Rust, J., Wedmann, S., Bradler, S. et al. (2013). Is Strudiella a Devonian insect? *Nature*, 494, E3E4.

Jacquelin, L., Desutter-Grandcolas, L., Chintauan-Marquier, I., Boistel, R., Zheng, D., Prokop, J., Nel, A. (2018). New insights on basivenal sclerites using 3D tools and homology of wing veins in Odonatoptera (Insecta). *Scientific Reports*, 8(238), 1–7.

Jouault, C., Legendre, F., Grandcolas, P., Nel, A. (2021). Revising dating estimates and the antiquity of eusociality in termites using the fossilized birth–death process. *Systematic Entomology*, 46, 592–610.

Jouault, C., Maréchal, A., Condamine, F.L., Wang, B., Nel, A., Legendre, F., Perrichot, V. (2022a). Including fossils in phylogeny: A glimpse into the evolution of the superfamily Evanioidea (Hymenoptera: Apocrita) under tipdating and the fossilized birth–death process. *Zoological Journal of the Linnean Society*, 194, 1396–1423.

Jouault, C., Nel, A., Perrichot, V., Legendre, F., Condamine, F.L. (2022b). Multiple drivers and lineage-specific insect extinctions during the Permo–Triassic. *Nature Communications*, 13(7512), 1–16.

Jouault, C., Nel, A., Legendre, F., Condamine, F.L. (2022c). Estimating the drivers of diversification of stoneflies through time and the limits of their fossil record. *Insect Systematics and Diversity*, 6(4), 1–14.

Khramov, A.V., Foraponova, T., Węgierek, P. (2023). The earliest pollen-loaded insects from the Lower Permian of Russia. *Biology Letters*, 19(20220523), 1–6.

Labandeira, C.C., Wilf, P., Johnson, K.R., Marsh, F. (2007). *Guide to Insect (and Other) Damage Types on Compressed Plant Fossils (Version 3.0 – Spring 2007)*. Smithsonian Institution, Washington.

Misof, B., Liu, S., Meusemann, K., Peters, R.S., Donath, A., Mayer, C., Frandsen, P.B., Ware, J., Flouri, T., Beutel, R.G. et al. (2014). Phylogenomics resolves the timing and pattern of insect evolution. *Science*, 346, 763–767.

Montagna, M., Tong, K.J., Magoga, G., Strada, L., Tintori, A., Ho, S.Y.W., Lo, N. (2019). Recalibration of the insect evolutionary time scale using Monte San Giorgio fossils suggests survival of key lineages through the End-Permian Extinction. *Proceedings of the Royal Society (B)*, 286(20191854), 1–9.

Nel, A. and Piney, B. (2023). The Odonatoptera: A clade that contains 99% of Odonata fossil diversity. In *Dragonflies and Damselflies*, 2nd edition, Cordoba-Aguilar, A., Beatty, C., Jason Bried, J. (eds). Oxford University Press, Oxford, 279–293.

Nel, A., Roques, P., Nel, P., Prokin, A.A., Bourgoin, T., Prokop, J., Szwedo, J., Azar, D., Desutter-Grandcolas, L., Wappler, T. et al. (2013). The earliest known holometabolous insects. *Nature*, 503, 257–261.

Nel, P., Bertrand, S., Nel, A. (2018). Diversification of insects since the Devonian: A new approach based on morphological disparity of mouthparts. *Scientific Reports*, 8(3516), 1–45.

Nel, A., Garrouste, R., Kaya, M., Licht, A., Legal, S., Coster, P. (2023a). The second oldest representative of the genus *Aeshna* (Odonata: Aeshnidae) found in the lowermost Oligocene of Luberon (France) and revealed by UV light. *Historical Biology*, 36(13) 1–5. doi: 10.1080/08912963.2022.2157274.

Nel, A., Wei, M., Niu, G., Garrouste, R., Jouault, C. (2023b). Description of a new fossil genus of conifer sawfly (Hymenoptera: Diprionidae) revealed by UV light. *Palaeoentomology*, 6(3), 313–320.

Peris, D., Janssen, K., Barthel, H.J., Bierbaum, G., Delclòs, X., Peñalver, E., Solorzano-Kraemer, M.M., Jordal, B.H., Rust, J. (2020). DNA from resin-embedded organisms: Past, present and future. *PLoS ONE*, 15(9), 1–21.

Ronquist, F., Klopfstein, S., Vilhelmsen, L., Schulmeister, S., Murray, D.L., Rasnitsyn, A.P. (2012). A total-evidence approach to dating with fossils, applied to the early radiation of the Hymenoptera. *Systematic Biology*, 61, 973–999.

Schubnel, T., Desutter-Grandcolas, L., Legendre, F., Prokop, J., Mazurier, A., Garrouste, R., Grandcolas, P., Nel, A. (2020). To be or not to be: Postcubital vein in insects revealed by microtomography. *Systematic Entomology*, 45, 327–336.

Sollai, G. and Solari, P. (2022). An overview of "Insect Biodiversity". *Diversity*, 14(134), 1–5.

Stork, N.E. (2018). How many species of insects and other terrestrial arthropods are there on Earth? *Annual Review of Entomology*, 63, 31–45.

Tafforeau, P., Boistel, R., Boller, E., Bravin, A., Brunet, M., Chaimanee, Y., Cloetens, P., Feist, M., Hoszowska, J., Jaeger, J.-J. et al. (2006). Applications of X-ray synchrotron microtomography for non-destructive 3D studies of paleontological specimens. *Applied Physics, Materials Science and Processing (A)*, 83, 195–202.

Tong, K.J., Duchêne, S., Ho, S.Y.W., Lo, N. (2015). Comment on "Phylogenomics resolves the timing and pattern of insect evolution". *Science*, 349(6247), 487.

Vea, I.M. and Grimaldi, D.A. (2016). Putting scales into evolutionary time: The divergence of major scale insect lineages (Hemiptera) predates the radiation of modern angiosperm hosts. *Scientific Reports*, 6(23487), 1–11.

Wang, Y., Engel, M.S., Rafael, J.A., Wu, H., Rédei, D., Xie, Q., Wang, G., Liu, X., Bu, W. (2016). Fossil record of stem groups employed in evaluating the chronogram of insects (Arthropoda: Hexapoda). *Scientific Reports*, 6(38939), 1–12.

Wappler, T., Currano, E.D., Wilf, P., Rust, J., Labandeira, C.C. (2009). No post-Cretaceous ecosystem depression in European forests? Rich insect-feeding damage on diverse middle Palaeocene plants, Menat, France. *Proceedings of the Royal Society B*, 276, 4271–4277.

9

X-ray Tomography of Crinoids: Morphological Diversity and Evolution Seen under a Different Light

9.1. X-ray microtomography

X-ray microtomography, also known as Computed Tomography (CT), is a noninvasive 3D imaging technique which allows us to see internal microstructures, notably, in natural history specimens. It allows users to observe section images of the object, and we can virtually dissect the specimens to produce 3D models of their anatomic structures. This technique is used to complement 2D observations (photography, electron microscopy, histology) for morphological analysis.

A CT scanner is made up of an X-ray source, a platform to hold the specimen being examined, and an X-ray detection system (see Figure 9.1). This device is used in conjunction with a computer to render the images. Tomographic acquisition consists of capturing a series of 2D radiographic images of an object while it completes a whole rotation around its vertical axis. A radiograph is a projection of the object onto a plane, corresponding to the number of X-rays reaching the detector. The contrast in radiography is created by variations in the X-ray absorption levels of the specimen's

Chapter written by Pablo MARTINEZ-SOARES, Nadia AMÉZIANE, Marta BELLATO, Julien BIGOT, Margot MATHIEU, Michel ROUX, Patricia WILLS and Marc ELÉAUME.

different structures. Thus, mineralized structures can be distinguished from soft tissues by their greater attenuation.

Using a series of radiographs, an algorithm creates a tomographic reconstruction of the object: a 3D matrix of voxels (a contraction of *volumetric pixels*) depicting the object, where each voxel represents a local attenuation coefficient. The size of a voxel in a tomographic image depends on the detector's characteristics (notably, the pixel size of the detector) and the magnification (distances between the source, the object and the detector).

A voxel is generally somewhere between 5 µm and 150 µm in laboratory equipment. The data from a tomographic acquisition are saved in the form of a stack of 2D images, representing the cross-sections of the volume in a given direction, often the horizontal plane.

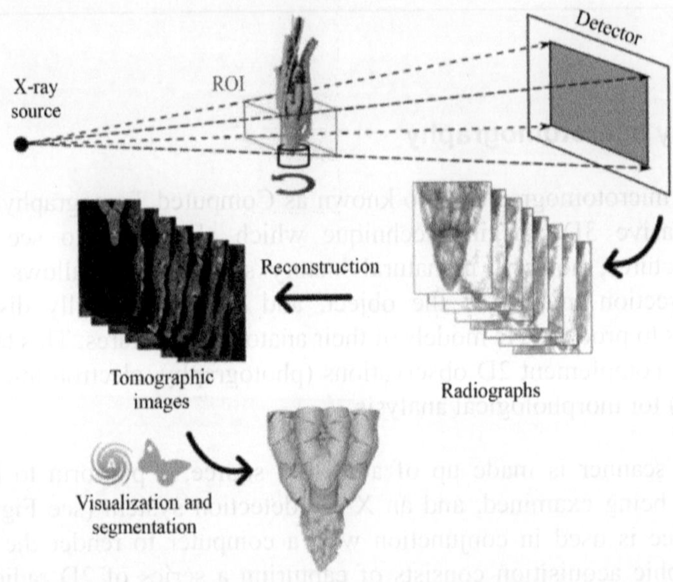

Figure 9.1. *Diagrammatic representation of the process of computed tomography, and the results of the reconstruction and segmentation steps. Here, we see the logos of the software packages Avizo (the big A) and Biomedisa (the blue butterfly), because they are greatly used in preparing our samples. ROI is Region of Interest. For a color version of this figure, see www.iste.co.uk/nicolas/biodiversity.zip*

9.2. The sample

The sample must be stable while the images are acquired, in order for the reconstruction algorithm to work properly. If the object moves between the first and last radiographs captured, the algorithm produces a dataset with a fuzzy effect, and the boundaries of objects appear blurred and doubled. The samples are therefore stabilized with foam. Those preserved in ethanol are first coated in a plastic film to keep them humid. Drying out could not only impact the preservation of the structures, but also potentially lead to a deformation of the object, causing blurring.

It is also possible to capture an image of a smaller area of the object (the Region of Interest: ROI) if it cannot fully fit within the field of view. In addition, certain systems allow the detector to be moved vertically or horizontally, thus enlarging the field of view in order to capture larger objects (up to a few tens of centimeters). By defining a specific ROI, we are able to obtain a better resolution compared to when scanning a whole object.

The composition of the sample may create constraints for X-ray tomography in terms of absorption-based contrast. For example, internal structures, which have the same density as their immediate surroundings, will be difficult to distinguish. Such is the case with certain fossils, where the mineralized specimen has the same density as the rock matrix surrounding it. Conversely, fossils, which contain metallic inclusions (e.g. containing pyrite in the specimen or ferrous nodules in the matrix), will present very high X-ray absorption, thus degrading the quality of the tomographic images. Soft tissues, such as muscles, have low X-ray absorption, and therefore appear with little contrast. Magnetic-resonance imaging (MRI) or contrast enhancers based on heavy elements may be used to better view these tissues on the reconstructed images.

9.3. Software

There are many available software packages for processing reconstructed images: some free and others commercial. Among others, we can point to those which our team has used and tested in recent years: Avizo (Thermo Fisher Scientific) and Mimics (Materialise NV), both of which are paid for, Dragonfly (Object Research Systems) which is free for noncommercial use, and 3Dslicer which is completely free and open source. We have also made ample use of the free application Biomedisa (Lösel et al. 2020). This application reduces the effort invested during the segmentation phase. Indeed, segmentation is a process which can be both time-consuming and error-prone. Instead of needing to segment each 2D image or perform a basic morphological interpolation, Biomedisa (and also Dragonfly) "intelligently" interpolates the images, using all of the information from the underlying image. Thus, we only need to segment around one image in twenty. However, the software will not be able to distinguish structures in an image if their boundary is not clearly defined. Note that increasing numbers of software packages are starting to implement artificial intelligence (AI) systems. For example, a neural network can be trained to segment new images, once we have fed it a set of pre-segmented images. These highly promising methods will probably soon help to speed up and automate image processing for 3D imaging.

9.4. X-ray study of crinoids

9.4.1. *Presentation and current issue with crinoids*

Crinoids are a class of echinoderms which emerged during the Lower Ordovician (some 470 mya). Today, these organisms are found in practically all seas and oceans, from coastal waters down to the abyss. Most are considered to be passive suspension feeders, more or less rheophilic, forming a filtration fan with their arms, pinnules and podia to capture particles suspended in the water column.

Their bodies are made up of a calyx, which includes the visceral mass, supported by a stalk, and extended by a crown of arms. The arms laterally support appendices known as pinnules. In common native speech, we distinguish between two morphotypes: *sea-lilies*, which keep their stalks in adulthood, and *feather-stars*, which lose their stalk after their post-larval stage. Crinoids have a calcitic endoskeleton comprising numerous ossicles: those in the stalk are called columnals; those in the arms are brachials; those in the pinnules are called pinnulars.

The study of the ontogeny of the articulations between brachials and between columnals (Améziane and Roux 2005; Améziane et al. 2021), as well as the first molecular phylogenies (Hemery et al. 2013; Rouse et al. 2013) turned the conventional classification of this clade on its head. In fact, numerous morphological characters used in crinoid systematics are cases of homoplasy. Therefore, we seek new, robust morphological characters, which will allow us to reevaluate the systematics of modern-day and fossil crinoids, and better understand how they have evolved. We seek to reconcile ontogenetic and molecular approaches in order to better comprehend the origin of today's species.

9.4.2. *From radiography to CT scans*

In the beginning, many studies were carried out using histological and/or paleohistological sections, notably for the study of the aboral nervous system. Then, it was the dawn of radiography that opened up new horizons. Roux (1977) developed this technique to show ontogenetic witnesses and growth zones of crinoid skeletons, which become visible thanks to variations in the density and microstructure of the calcitic lattice, called stereom. However, the overlapping of different microstructures within the ossicles limits the precision of the images, and makes them difficult to interpret.

To our knowledge, the first use of a CT scan on a crinoid was presented by Aschauer et al. in 2006. These authors compared the reconstruction of internal structures in crinoids using various methods of histological coating and a CT scan. They conclude that CT scans are a relevant tool to better understand the general anatomy of crinoids. However, in the years that have followed, there have been few studies which have employed this technique, probably limited by the monetary cost, the necessary computer resources and a lack of a clear objective justifying the use of this method. To date, the first and only large-scale comparative anatomy study using CT scans on crinoids was carried out by Saulsbury (2019).

9.4.3. *Crinoids and CT scanning at the MNHN: various ongoing projects*

9.4.3.1. *Study of fossil crinoids: example of Bourgueticrinina*

Bourgueticrinina are a group of crinoids with a xenomorphic stalk, that is, they are made up of columnals whose morphology varies significantly depending on where in the peduncle they are located. This variation is linked to changes in the means of articulation as the organism grows. Molecular phylogenetic reconstructions (Hemery et al. 2013; Rouse et al. 2013) place Bourgueticrinina in the Comatulida; an order which originally contained only feather-stars (crinoids which do not have a stalk in their adult stage). In addition, Hemery et al. (2013) showed that a family of feather-stars, the Atelecrinidae, is nested within the Bourgueticrinina clade.

Furthermore, keeping a xenomorphic stalk until adulthood is a pedomorphic character that Bourgueticrinina share with other clades. Exploration of the morphological characteristics of the arms showed that, with the exception of Phrynocrinidae, Bourgueticrinina (including Atelecrinidae) have special structures present on the proximal brachials, known as *knobby processes*. This may be a synapomorphy characterizing this clade. However, to date, no knobby processes have been identified in fossilized Bourgueticrinina. Other morphological characteristics therefore need to be found in order to form the connection between fossil species and those alive today.

Marls (a type of sedimentary rock) from the Eocene of Possagno, north-eastern Italy, could provide new clues. In these marls, the skeleton preservation of *Paraconocrinus pyriformis* (Bourgueticrinina, Rhizocrinidae) is remarkable, with the ossicles still partially connected (see Figure 9.2). In addition, the knobby processes can be preserved, despite their fragility.

Thanks to CT scans, we can observe connected stalk fragments, meaning it is possible to analyze the shape of the columnals along the stalk. However, the first sample contained no connected arms. Nevertheless, we were able to find several isolated brachials. The calices, which are present in the sample, could be segmented and used to study the intraspecific variation of the ligamentary areas, the interradial ridges, and the aboral nervous system.

Proper conservation and quality of the 3D reconstruction of the skeletal elements enable us to take classic morphometric measurements of the columnals and brachials. In future marl samples, we hope to be able to find the proximal part of the stalk still attached to a calyx, and parts of the arms fully or partly preserved.

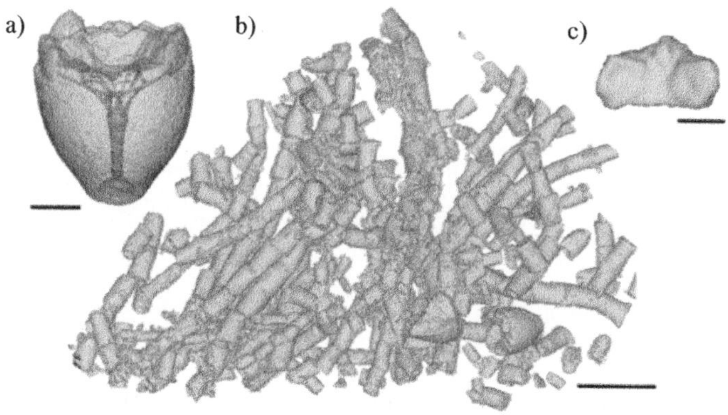

Figure 9.2. *Reconstructions of fossils found in marl in Possagno (Italy): a) Calyx of* Paraconocrinus pyriformis, *showing the aboral nervous system visible in transparency (dark gray); b) reconstruction of all of the ossicles in the marl in the original position in which they were buried; c) oral view of a possible axillary of* P. pyriformis. *The scales show 1 mm, 1 cm and 0.5 cm, respectively*

9.4.3.2. *Study of ontogenesis: the example of* Holopus alidis

The Holopodidae are a family of crinoids comprising two genera, six living species and eight fossil species having been described. They do not have a stalk, and their calyx is directly attached to a hard substrate. They roll up their arms over themselves to form a "casing", covering not only the distal part of their arms, but also their mouth and tegmen. These organisms are mostly found on undersea overhangs, seamounts, and in undersea caves between 200 and 800 meters depth.

To better understand the evolution of Holopodidae, it is necessary to determine the homologous characters of this group. Using CT scanning for this analysis prevents damaging the invaluable specimens preserved in the collections of the MNHN, and provides an exact 3D depiction of the position of each brachial along the arms. In particular, the brachials involved in forming the "casing". This approach enables us to measure the skeletal components, including elements which are inaccessible in external morphology, and compare individuals. These measurements allow us to describe changes in size and shape of the ossicles within an individual and from one individual to another, be it in an ontogenetic or phylogenetic context.

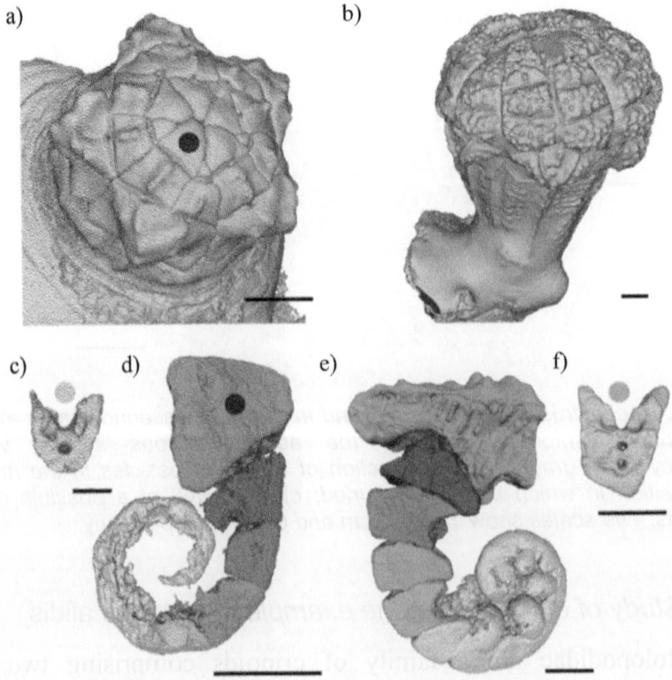

Figure 9.3. *Comparison of a juvenile and an adult specimen of* Holopus alidis. *For a color version of this figure, see www.iste.co.uk/nicolas/biodiversity.zip*

COMMENT ON FIGURE 9.3.– *a) Juvenile, three-quarter view, showing the surface of its casing; b) adult, lateral view; c) juvenile, proximal articular face of a brachial in the distal part of an arm; d) juvenile, lateral view of the keystone brachial and the distal part of an arm; e) adult, lateral view of the keystone brachial and the distal part of an arm; f) adult, proximal articular face of a brachial in the distal part. The scales indicate 1 mm except for (c) (0.25 mm). Each color represents the same ossicle on the various figures.*

In order to properly understand the variation of the morphological characters over geological timescales, it is important to have a solid comprehension of the variation at individual level, and its ontogeny in extant species. For this reason, we scanned the rare juvenile specimens in the MNHN's collections. In *Holopus alidis*, both juvenile and adult show a hermetic casing (see Figure 9.3). Following segmentation, we can observe three major differences between the brachials at the two stages: in the juvenile, i) the brachials in the keystone position do not have structures allowing them to interlock with neighboring ones; ii) the brachials in the keystone position do not have a developed distal outgrowth; iii) the brachials in the distal part of the arms exhibit a rounded rather than angular cross-sectional shape. In juveniles, as they are not specialized, the brachials in the keystone position could turn into subrectangular brachials typical of the proximal part of *Holopus* arms. We can then suppose that once the animals reach a certain size, the fourth or fifth brachial in the keystone position becomes specialized and develops a long distal outgrowth. This outgrowth can fill the spaces which form between the arms as the crinoid continues to grow. This hypothesis is supported by the growth lines visible on the scans (see Figures 9.4c and 9.4e). In addition, the growth of the distal part of the different arms within the casing brings them into contact. To maximize the use of the limited space available, they develop the angular shape which we typically see in adults (see Figure 9.4b).

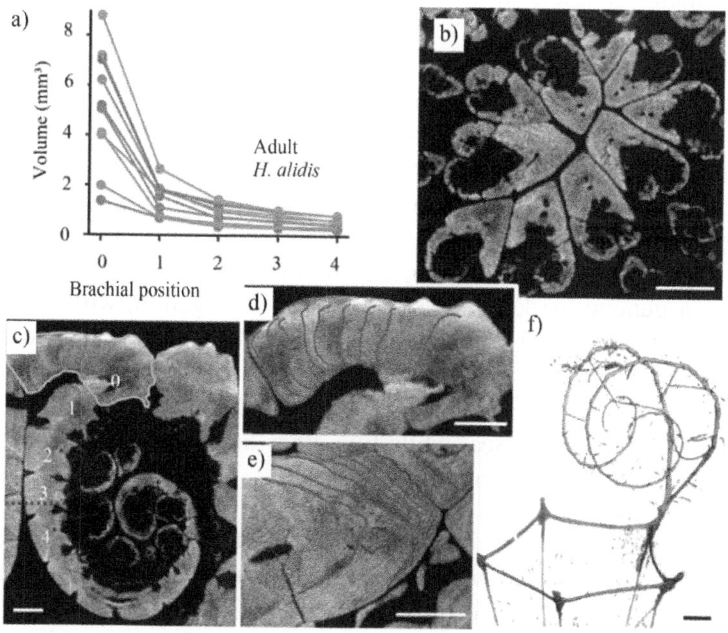

Figure 9.4. *Observations in an adult* H. Alidis *exploiting different capabilities of CT scans. For a color version of this figure, see www.iste.co.uk/nicolas/biodiversity.zip*

COMMENT ON FIGURE 9.4.– *a) Graphic showing the decrease in brachial volume (obtained from their segmentation) within the animal's casing.*

Position zero corresponds to the keystone brachial. Each color corresponds to one of the ten arms; b) transverse tomography image within the casing showing the arrangement of the brachials in the distal part; c) sagittal tomography image across the distal part of an arm. Contours of the keystone brachial in yellow. The numbers correspond to the position of the brachials as defined in (a). The dotted line corresponds to the level of the sagittal slice of (b); d) close-up view of the keystone brachial from (c); e) transverse tomography image across a keystone brachial. The red lines indicate growth lines; f) reconstruction of the aboral nervous system in the calyx (dark gray) and two arms (light gray). The scales indicate 1 mm.

9.4.3.3. *Morphological diversity study: the example of Notocrinus*

Notocrinus is a genus of feather-star endemic to the Southern Ocean, with a circumpolar distribution. They are brooding crinoids, having a pouch at the base of their pinnules, in which they shelter their larvae. The literature distinguishes two species: *N. virilis* and *N. mortenseni*. *N. virilis* is characterized notably by a high, conical centrodorsal, with long cirri arranged in columns. *N. mortenseni* has a low centrodorsal and short cirri with no particular arrangement on the centrodorsal. Phylogenetic reconstructions, haplotype networks and species delimitation methods show that these two species are complexes of species. Thus, there are thought to be at least two cryptic species for *N. mortenseni*, and at least three for *N. virilis*. In the case of *N. mortenseni*, the use of cytochrome c oxidase subunit I (COI) as a genetic marker shows geographic segregation, with at least one species in the Weddell Sea and the Scotia Arc, and a second species collected on the Burdwood Bank. In the case of *N. virilis*, the results are more complex. One line seems to be restricted to the Ross Sea, Dumont d'Urville Sea and Davis Sea, while two other lines seem to be restricted to the Weddell Sea and the Scotia Arc.

With CT scans, we are able to see the major morphological differences between *N. virilis* and *N. mortenseni* (see Figure 9.5). Thus, five celomic cavities are visible in the centrodorsals of *N. virilis*; these structures are absent in *N. mortenseni*. Similarly, the basals are well developed in *N. virilis*, forming a clear circle, but are reduced or absent in *N. mortenseni*.

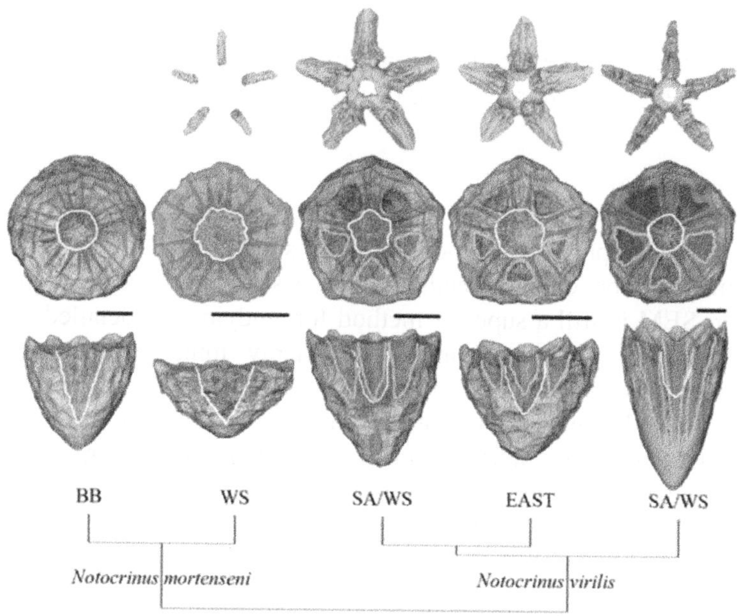

Figure 9.5. *Scans of different lineages of* Notocrinus *assessed with molecular methods. For a color version of this figure, see www.iste.co.uk/nicolas/biodiversity.zip*

COMMENT ON FIGURE 9.5.– *First row: circle of basals. Second row: oral view of segmented centrodorsals. Third row: lateral view of centrodorsals. For each specimen, the white line marks the edge of the central cavity, and the yellow lines delimit three of the five celomic cavities. The scale shows 2mm. BB: Burdwood Bank; WS: Weddell Sea; SA: Scotia Arc; EAST: East Antarctica (Ross Sea, Dumont d'Urville Sea, and Davis Sea).*

Morphological differences between cryptic species in each complex can also be demonstrated. For larger *N. mortenseni*, the central cavity is proportionally narrower, less deep and has cirri that are less wide in the individual from the Burdwood Bank. In addition, the specimen from the Weddell Sea has small, underdeveloped basal, while the Burdwood Bank specimen seems to have none at all. In *N. virilis*, the line from Eastern Antarctica has shorter celomic cavities than the central cavity. In the Weddell Sea, one line has both structures of around the same depth, whereas in the other, the celomic cavities are twice as deep.

The importance of these morphological differences and the barcode gap could justify at least the creation of multiple species within each complex, if not the creation of an entirely separate genus.

9.5. Conclusion

X-ray tomography is a method perfectly suited to the study of crinoids, complementing conventional methods such as scanning electron microscopy. However, SEM is still a superior method for studying the detailed structure of the stereom. As it is noninvasive, X-ray tomography lends itself particularly well to the study of unique crinoids (such as type specimens), or rare ones (e.g. the specimens collected in environments that are difficult to access). It is also particularly useful in analyzing fossils within their rock matrix, or internal structures such as the arms of Holopodidae or the celomic cavities of *Notocrinus*. Tomographic images also allow us to observe ossicles in cross-sections, revealing the growth lines without the problem of overlap that occurs in radiography. 3D reconstructions can be used to measure conventional morphometric variables, and also to measure volumes and use 3D geometric morphometric methods. The segmentation of structures is still a time-consuming task, but smart interpolation and artificial intelligence to segment whole samples have already made this process significantly easier. It is highly likely that future segmentation software will make this technique even easier to use.

9.6. References

Améziane, N. and Roux, M. (2005). Environmental control versus phylogenic fingerprint in ontogeny: The example of the development of the stalk in the genus *Guillecrinus* (stalked crinoids, Echinodermata). *J. Nat. Hist.*, 39(30), 2815–2860.

Améziane, N., Eléaume, M., Roux, M. (2021). Ontogeny of non-muscular brachial articulations in Balanocrininae (Echinodermata, Crinoidea): Iterative trajectories or phylogenetic significance? *Zoomorphology*, 140(1), 47–67.

Aschauer, B., Heinzeller, T., Weinert, P. (2006). Almost within grasp: Crinoid organs rendered 3-dimensionally. In *Echinoderms Durham: Proceedings of the 12th International Echinoderm Conference*. CRC Press, Leiden, 9–14.

Hemery, L.G., Roux, M., Ameziane, N., Eleaume, M. (2013). High-resolution crinoid phyletic inter-relationships derived from molecular data. *Cah. Biol. Mar.*, 54, 511–523.

Lösel, P.D., van de Kamp, T., Jayme, A., Ershov, A., Faragó, T., Pichler, O., Tan Jerome, N., Aadepu, N., Bremer, S., Chilingaryan, S.A. et al. (2020). Introducing Biomedisa as an open-source online platform for biomedical image segmentation. *Nat. Commun.*, 11(1), 5577.

Rouse, G.W., Jermiin, L.S., Wilson, N.G., Eeckhaut, I., Lanterbecq, D., Oji, T., Young, C.M., Browning, T., Cisternas, P., Helgen, L.E. et al. (2013). Fixed, free, and fixed: The fickle phylogeny of extant Crinoidea (Echinodermata) and their Permian – Triassic origin. *Mol. Phylogenet. Evol.*, 66(1), 161–181.

Roux, M. (1977). Les Bourgueticrinina (Crinoidea) recueillis par la 'Thalassa' dans le golfe de Gascogne : anatomie comparée des pédoncules et systématique. *Bulletin du Muséum national d'histoire naturelle*, 3(426), 1–83.

Saulsbury, J. (2019). Crinoid respiration and the distribution of energetic strategies among marine invertebrates. *Biol. J. Linn. Soc. Lond.*, 129(1), 244–258.

10

Conceptual and Methodological Foundations of Integrative Taxonomy

10.1. Introduction

Each year, over 20,000 new species are described, by some 30,000–40,000 taxonomists, both professional and amateur. This equates to approximately one every 30 minutes (Miralles et al. 2020). Though the inventory of the living world currently totals nearly 2.1 million species described, this project – which has been ongoing for almost 300 years – is far from complete. Recent estimations, although they vary, agree that the planet is home to at least ten million eucaryotic species (Larsen et al. 2017). Thus, at the current rate of description, it would take several centuries to complete the task (Costello et al. 2013a). The sixth mass extinction event is only just beginning, and we simply no longer have the time. Taxonomy research must quickly adapt its practices to significantly speed up the inventory of the living world, but without compromising the robustness of the hypotheses on the species described. Recent technological and conceptual advances mean that it is now possible to speed up the process, and their combined power must be used to this effect. Our real-time understanding of the coming global crisis, and our ability to address that crisis, depend on it (May and Harvey 2009).

There are many methodological challenges facing taxonomy, some quantitative (how to accelerate the process of data and knowledge production, and facilitate the dissemination of this knowledge); and some

Chapter written by Aurélien MIRALLES and Nicolas PUILLANDRE.

qualitative (advancing more robust hypotheses on species, which will therefore be more stable over time). These challenges include:

– the transfer and appropriation, by the community, of concepts and methods stemming from evolutionary biology;

– the standardization of decision-making processes;

– the integration of new types of data (genomes, transcriptomes, CT scans, etc.);

– the development of suitable digital tools (interoperable and specimen-centered data, tools devoted to delimitation in an integrative context and assessment of the robustness of the taxonomic hypotheses).

In this chapter, devoted to recent developments in alpha taxonomy, we present what taxonomy is, and certain key concepts associated with it. A brief overview of the history of the discipline will illustrate the steps which have had a lasting impact on the discipline, and helped move it forward. This chronological perspective should give readers a better understanding of the scope of the paradigm shift, which some refer to as a "taxonomic renaissance", that has been occurring in the discipline since the turn of the millennium.

10.2. A multifaceted discipline

10.2.1. *Defining taxonomy*

Taxonomy is the oldest of the natural sciences. It aims to reduce the staggering structural complexity of living organisms to a simplified, categorical, unambiguous communication system, universally reusable by the other disciplines of biology, and whose elementary unit is the species category. Taxonomy constructs, defines, names and classifies the different interlinked categories of organisms which are collectively referred to as taxa (Miralles 2021).

Two branches of taxonomic practice, working at different scales and with very different goals, can be identified (Mayr and Ashlock 1991). To begin with, the task of inventorying the living world is addressed by alpha taxonomy (or microtaxonomy). It is this branch, which involves identifying and characterizing the elementary categorical units (living or extinct species), which will be discussed in this chapter. On the other hand, the classification of the living world is the task of beta taxonomy (or macrotaxonomy), consisting of classifying those same species units by assigning each of them to broader ranks, themselves arranged hierarchically.

Thus, the original purpose of taxonomy was to produce a simplified representation of the structure of the living world and its exploitability, through an effective communication system. In that respect, it differs from systematics as it is a more inclusive concept which encapsulates taxonomy. For its part, the purpose of systematics is to endow that structure with meaning, by tracing its evolutionary history or revealing the mechanisms by which it diversifies in space and over time. However, in contemporary practice, the distinction between these two "aims" is becoming less clear, as taxonomy adopts the interpretative and methodological framework of evolutionary biology and population genetics. Thus, β taxonomy appears to be ceding its place to phylogenetic classification of the living world, while year on year, α taxonomy is adopting integrative, evolutionary and population-based approaches.

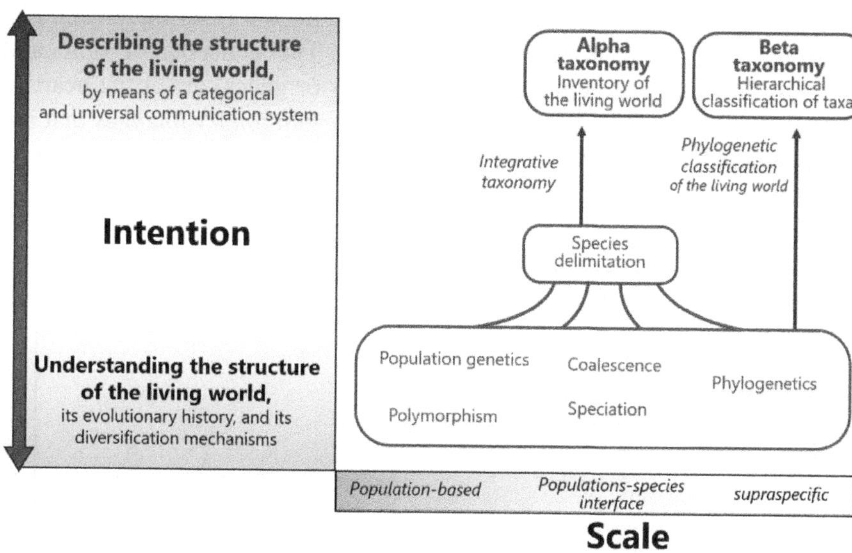

Figure 10.1. *The main task of alpha taxonomy is species delimitation. Increasingly, it achieves this goal using an approach from integrative taxonomy, which itself draws on the concepts and tools of population genetics and phylogeny. For a color version of this figure, see www.iste.co.uk/nicolas/biodiversity.zip*

COMMENT ON FIGURE 10.1.– *The objective of beta taxonomy is to produce a hierarchical classification of the living world, in which each species is included in interlocking groups, increasingly inclusive. This classification is now based on phylogeny.*

10.2.2. *The two tasks of contemporary α taxonomy*

Two major tasks in α-taxonomic practice can be distinguished: one fundamental, and the other applied. The first, the *fundamental* activity, is research, and thus the production of new scientific knowledge. It draws on an evolutionary, population-based and constructivist approach, and consists of an empirical study of the structure – more or less marked, but objective – of the "spectrum of variations of the living world", based on the study of individuals. This spectrum, produced from the evolutionary tree, objectively combines two apparently contradictory features: *genealogical continuity* is expressed on the time axis, but little by little, it leads to *reproductive discontinuity* which can be observed in the present. It is in bringing these two axes together that contemporary taxonomy differs from the essentialist and fixist approaches from which it sprang. The concepts of time and variability (diversity of populations, individuals or genes) are at the heart of the disciples, and the species are assimilated to *evolutionary lineages* that are independent of one another (de Queiroz 1998; Samadi and Barberousse 2006).

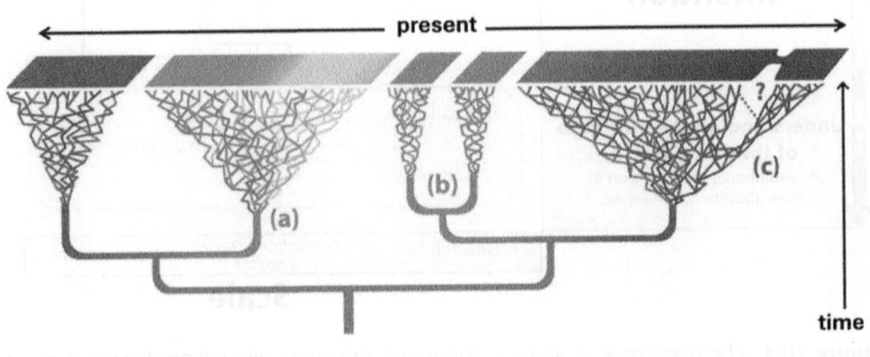

Figure 10.2. *Reconciling the discontinuity present in the living world with its continuity observed over time. For a color version of this figure, see www.iste.co.uk/nicolas/biodiversity.zip*

COMMENT ON FIGURE 10.2.– *In practice, the issues involved in taxonomic work consist essentially of not considering as distinct species those populations whose visible divergence is, in reality, merely intraspecies polymorphism (a); and conversely, differentiating cryptic species (i.e. those which appear identical, despite having evolved independently of one another (b)). The case of more recent speciations is trickier to handle (c), given the frequent impossibility of knowing for certain whether the divergence process is complete (permanent) or ongoing (more or less reversible). In this scenario, the decision made by a taxonomist in view of the data available at time* t *will contain an inevitable element of uncertainty and arbitrariness. However, the decision should be justified as far as possible (modified from Miralles 2024).*

The second task, which is applied, corresponds to the standardization of taxonomic knowledge and is regulated by nomenclature codes. This takes place as a second step, and provides other scientific disciplines – and indeed society in general – with a universal communication system. This system can be used to name the myriad species currently recognized, but also to identify them (assessment and identification tools). As the fundamental research mentioned above advances, our knowledge becomes refined, and the hypotheses on species evolve: a species hypothesis put forward at time *t* depends on the data available at that time, and like any scientific hypothesis, can be modified on the basis of new data. This system, which is unstable by nature, therefore needs to be updated periodically, by taxonomic changes published in scientific journals, in the form of nomenclatural acts or opinions. The most iconic of these acts, the description of new taxonomic species, characterizes and officially records the new concepts they are named after: firstly, the new species hypothesis is defended, setting out the biological arguments previously compiled in support of it (this is, of course, the product of the fundamental research aspect). The "description" of the new species is regulatory, and involves applying a procedure defined by the international nomenclature codes. Its application ensures that the living world is inventoried in a standardized way. Also, amongst other things, it helps prevent total chaos of nomenclature, or possible conflicts between researchers claiming credit for the same discovery. In zoology, a unique scientific binomen combining a genus name (e.g. *Pongo*) with a specific label created for the occasion (e.g. *tapanuliensis*) would be attached to the species, and a specimen deposited in a reference collection denoted as the holotype for that species. This material object is the indisputable anchor point in the real world, forming the connection between these two thought products: the binomen (a term) and the species hypothesis (a concept). All researchers, all over the world, can examine this universal and timeless reference, notably in order to discover exactly what a name is attached to. Without these "Rosetta Stones", our capacity to communicate about the millions of species inhabiting the planet would be sorely hindered.

These two tasks, the fundamental and the applied, are thus different in nature, but nevertheless closely connected. The connection between them could be summarized simply as follows: any description of a nominal species is in keeping with its delimitation, as far as possible in an integrative context; and conversely, should any hypothesis subsequently be amended or disproven, the nomenclature must be updated (synonymy established, a new species described, etc.).

10.3. A brief history of the taxonomic paradigm, from Linnaeus to the modern world

10.3.1. *Origins*

Human societies have undoubtedly always needed to distinguish the different forms of organisms and give them distinct names. These diverse parataxonomies vary along with languages and cultures; in contrast, Linnaean nomenclature offers a universal, standardized and unambiguous system for naming living or extinct species. *Carl von Linnaeus* was the founder and initiator of the binomial classification system which has spread to universal acceptance since the tenth edition of *Systema Naturae* (1758). The taxonomy of the time was based on a fixist and essentialist paradigm, as Linnaeus' species, stemming from Genesis, conformed to the Platonic archetype: intraspecies polymorphism, seen as simple variations on the same theme, was deemed to have little relevance.

10.3.2. *Evolution(s)*

A century later, Charles Darwin's (1859) introduction of the theory of evolution by natural selection revolutionized our relationship with the diversity of the living world. At the time of Darwin's publication, it had very little impact on the classification methods used to organize the higher orders of organisms, and absolutely none on α-taxonomic practices. It was not until the mid-20th century that two major schools of thought emerged, enabling taxonomy sensu lato to shed the shackles of its history. The first was the synthetic theory of evolution, forming the link between Darwinian selection and Mendelian heredity, establishing the basis on population genetics. Within this school of thought, Theodosius Dobzhansky and Ernst Mayr focused particularly on the mechanisms of species formation (speciation), and developed the biological (or population-based) concept of the species (see, for example, Dobzhansky 1937; Mayr 1942). Secondly, around the same time, Willi Hennig drew upon the notion of descendancy with modification to lay down the theoretical bases of phylogenetic reconstruction (Hennig 1950). Thus, nearly a century after the publication of Darwin's opus, the outlines of a general theoretic evolutionist framework were sketched out: one which is consistent to describe the structure of the living world at its different levels.

The final barriers to the concrete application of Darwinian principles were partly technological. A number of these barriers were overcome in the 1990s, notably due to expanding access to bio-informatics and molecular biology. From that point on, the number of studies relating to phylogenetics, population genetics or phylogeography has grown year on year, documenting the history of the living world with a degree of informativity that would have been difficult to imagine only a decade earlier. It was also during this period that there was a perceptible split between the "alpha" and "beta" components of taxonomy: whereas the *classification of the living world* was able to be rapidly reinvented, taking advantage of methodological and technological advances, and attract the attention of a number of evolutionary biologists, the *inventory of the living world* is far from receiving such attention. At the time, the α-taxonomic community appeared to have difficulties in adopting the Darwinian paradigm and modern technologies (Wiens and Servedio 2000; Camargo and Sites 2013).

In the early 2000s, the study of scenarios relating to the evolutionary history of a given group was often based on the species hypotheses as available in the literature, without seeking to verify them. The evolutionary principles and methods are still not widely mobilized in advance of the study (i.e. at the time of the definition or consolidation of the species hypotheses). While for some people, this seems to stem from the methodological and conceptual conservatism or a community that is too attached to its practices, and insufficiently open to other disciplines, for others, such two-speed systematics seems to suggest a lack of recognition and resources, attributable to disaffection for the discipline within the life sciences. Still other people would say that the problem could stem from a large proportion of amateur researchers, generally less qualified in evolutionary biology and/or with limited research resources (see, for example, Gewin 2002; Wheeler 2004; Wheeler et al. 2004; Wilson 2004; Will et al. 2005; Padial and De la Riva 2006; Garnett and Christidis 2017). Each of these opinions throws up legitimate questions, and also is indicative of the profound identity crisis which the discipline was experiencing at the time. Yet beyond these sociological considerations, we must not overlook the applicability (and ultimately, transferability) of evolutionary concepts to α-taxonomic practice, in order to truly appreciate the scale of the difficulties of that time. What if the supposed methodological delay then suffered by α taxonomy simply reflected a lack of operational solutions that could be directly used by the community? The efforts made to resolve this problem, and the barriers they helped to overcome at the dawn of the 21st century, strongly suggest that this is the case.

10.3.3. *Renaissance*

The 2000s marked a crucial turning point in the history of α taxonomy. Helped by the rise of new technologies, a wave of reformation, solidly evolutionary and described by certain authors as a *taxonomic renaissance*, reshaped α-taxonomic practice at various levels (Mallet and Wilmott 2003; Sites and Marshall 2003; Miller 2007; Wiens 2007; Padial et al. 2010). It hinges on three key concepts.

The concept of a species-lineage: unifying the diverse concepts of species. From a strictly applied point of view, the respective impacts of the different (evolutionary) concepts of a species formulated over the course of the 20th century ultimately proved fairly limited for taxonomic practice. To look only at the most popular as an example, the *biological concept of a species* even seems to have created a methodological impasse for the discipline. This is so that, firstly, because the existence of a number of particular reproductive phenomena (for example, lines with strictly asexual reproduction, as in the cases of thelytokous parthenogenesis) it demonstrates that it is not universally *applicable*. Secondly, it fails to account for a number of intermediary situations between full interfertility and strict reproductive isolation: various divergent populations may be reciprocally interfertile, though they exhibit alteration, to a greater or lesser degree, of this potential (example of hybrid depression within the same species, with clinal variation forming a ring species, occasional cases of introgression between supposedly distinct species, etc.). Furthermore, as it is not testable (or testable only with great difficulty) in the majority of research situations in taxonomy, the concept is quite simply not functional.

In 1998, Kevin de Queiroz proposed the *general lineage concept of species*, based on the principle that there is a quasi-consensus on the fact that the species correspond to evolutionary lineages in populational level whose trajectories are definitively independent. The benefit of this concept lies, in particular, in the reconciliation that it provides between the various preexisting concepts, by offering a general theory. Indeed, whilst several of these "recent" concepts seem to agree broadly on the nature of a species, their different attempts at formalization exhibit inconsistencies, either internally or with one another (for an overview of the different evolutionary concepts of a species, see de Queiroz 1998). Ultimately, all of them fail to reflect what a species is, through a universal concept. This stems from the fact that none of these concepts *are* actually concepts in the true sense. Rather, they correspond to the *properties* of a species. All appear to be assimilated to indicators that point to the independence of their evolutionary trajectories. Monophyly, interfertility, intraspecies recognition, morphological uniformity, ecologic and behavioral uniformity, etc., help *demonstrate (the barriers between) species*, and thus constitute relevant delimitation criteria. However, these various criteria gradually emerge through the process of speciation and the order in which they appear varies from one situation to another. In particular, in the cases of recent divergences, it is therefore expected that these different criteria (formerly concepts) will form a mosaic, and that because of this disconnect, they will clash. Thus, they all individually fail to define *what a species is* from an ontological point of view (for an approach which is critical of Queiroz's, also see Seifert 2020).

Beyond the theoretical and unifying value of this dissociation between the *concept of a species* and the *delimitation criteria* (two concepts which had, up until that point, been broadly confused with each other), it is in the operational aspect that the true benefit of this new conception lies (also see

Sites and Marshall 2004; de Queiroz 2007): its flexibility allows taxonomists to test various criteria (without having to choose one to the detriment of others), contrast them, and combine them, being able to identify the evolutionary processes which lie behind their incongruence.

The notion of "integration": multiplying and combining information sources. For nearly 250 years, taxonomy was almost exclusively morphological. Following the general concept of species-lineages, the idea gradually gained traction that it needed to open up to all types of data that could demonstrate divergences between populations, all of which could potentially support or refute a speciation hypothesis. At the turn of the century, increasingly affordable sequencing techniques and the development of numerous bio-informatic tools facilitated the emergence of molecular α taxonomy. Initially, the data produced consisted mainly of mitochondrial DNA sequences, which are easy to amplify and highly informative due to its high rate of evolution. The *DNA barcoding* initiative (Hebert 2003), for example, was an ambitious program to sequence the same fragment of mtDNA (CO1) in order to construct a standardized genetic database, allowing for molecular identifications of the living world (also see Vogler and Monaghan 2006). Very soon, barcoding also proved to be a promising technique for preliminary exploration of biodiversity and the demonstration of potential new species (DeSalle et al. 2005; Fouquet et al. 2007).

Whilst the emergence, in 2005, of the concept of integrative taxonomy (Dayrat 2005; Will et al. 2005) helped highlight the importance of molecular tools, the term more generally denotes an approach that takes account of all relevant data (e.g. morphological, molecular, ecological, behavioral, etc.) and their interpretation in the light of evolutionary biology (for example, see Rissler and Apodaca 2007; Vieites et al. 2009; Melville et al. 2019). Whilst the ideas advocated may seem trivial to some, this ignores the fact that so-called integrative studies – those which exploit and combine at least two independent approaches – are still in the minority in taxonomy, even today. For example, the vast majority new species of plants or insects (accounting for three quarters of eucaryotic diversity) are still described solely on the basis of their morphological characteristics (Miralles et al. 2020)!

We come now to the concept of "species delimitation": rationalizing, formalizing and automating taxonomic inferences. Favored by the rise of molecular data, which yield more information but are also more complex to process, the idea of objectively reexamining the making of taxonomic decisions gained ground in the early 21st century. It notably illustrates the desire to break free, as far as possible, from subjective assessments such as morphological or molecular divergences deemed by an expert to be "sufficient". The development of clearly formalized, repeatable protocols based on biological and evolutionary principles, allowed for a methodological and epistemological leap forward, reducing the subjectivity in α-taxonomic inferences. Multiple approaches – first manual and subsequently automated – would rationalize taxonomists' work (e.g. GMYC, Monaghan et al. 2009; BPP, Yang and Rannala 2010; ABGD, Puillandre et al. 2012; PTP, Kapli et al. 2016; LIMES, Ducasse et al. 2020; ASAP, Puillandre et al. 2020; and iTaxoTools, Vences et al. 2021). These tools, based on highly diverse approaches, allow for: 1) better formalization of the decision-making process (hierarchical ranking of criteria based on their relevance, be it theoretical or empirical) (Pante et al. 2015), a rational definition of a quantitative or qualitative threshold beyond which a species hypothesis can be considered sufficiently well supported to be described; 2) the repeatability of analyses (identically on an enriched dataset, or else using alternative analytical approaches on the same dataset); 3) easier transfer of methods drawn from other disciplines (population genetics, morphometrics, phylogenetics, coalescent theory); 4) quantification of the degrees of support for speciation hypotheses; and 5) the demonstration of potential incongruence between different datasets.

Together, these three concepts, which stem from one another – *species-lineage*, *integration* and *delimitation* – have contributed to a gradual transition to a new paradigm within the discipline: up to that point, taxonomists' work focused on patterns (evidentiating the species entities); since then, its primary aim has been to demonstrate the existence of *processes* behind those patterns (speciation events).

10.4. Taxonomy of tomorrow: issues and prospects

The priority issue which 21st-century taxonomy must face – because it is from this issue that all others stem – clearly lies in speeding up the process of inventory of the living world. Aside from the obvious and thorny question of the resources devoted to the discipline, three "methodological" levers are currently experiencing promising developments:

i) Producing and sharing raw taxonomic data faster: this necessitates continuing with efforts to increase the production, processing and digital dissemination of data (this touches on the concepts of *cyber-taxonomy* and the *extended specimen network*), automating such tedious and repetitive tasks as can be automated, at all levels of the research process, from data harvesting to the publication of an article (Miralles et al. 2020).

ii) Producing more robust taxonomic knowledge: evidently, speeding up the inventory of the living world must not lead us to compromise on the quality of the hypotheses which flow from it. Using overly inflationist methods, on the pretext of accelerating the inventory process, would be counterproductive (the reviewing tasks needed to correct these excesses are, in themselves, time-consuming and a far less attractive prospect, because they are less valued). It would also lead to unnecessary instability in the nomenclature, and be damaging both to other fields of research and to conservation efforts (for example, see Isaac et al. 2004; Meiri and Mace 2007). A "good species hypothesis" is, put simply, a hypothesis which remains unchanged over time despite the publication of new taxonomic research. In particular, this robustness must come from the cross-referencing of different datasets and types of analysis, from diversified and representative sampling of specimens, and from objective and repeatable taxonomic decisions (here, we can refer to the concept of integrative taxonomy (see Padial et al. 2010)). For this reason, qualitative and quantitative improvements should not be seen as antagonistic, but as complementary, because ultimately, both depend on an increase in raw-data production. If the correct balance between these two requirements is struck, the methods should drive each other forward.

iii) Popularizing the use of these new methods: whilst there is the evident need to move the discipline forward by improving the available methods and tools, or developing new ones, it is also sensible to make these resources available to as many taxonomists as possible. It is important not to ignore the major disparities within the discipline, and the risk of the two-speed taxonomic practice producing a widening gap over time. For example, α-taxonomists studying invertebrates describe nearly three times more species than those working on vertebrates, but only a minority make use of recently developed techniques such as molecular biology or morphometrics (Tancoigne et al. 2011; Miralles et al. 2020). One solution to reduce this gap (which is conceptual, methodological and cultural) notably involves making suitable tools and services available to the taxonomy community. In order to be more widely adopted, they must cater to the needs identified by practitioners themselves, be accessible to all, simple and user-friendly.

If used in conjunction, these three levers should open the door to the essential "industrialization" of the process of inventory of the living world, it is also important to clarify to prevent any misunderstandings: these advances in terms of automation are not – and will never be – an acceptable substitute for the expertise of taxonomic researchers or the use of physical collections. On the contrary, this progress will free researchers from numerous tedious, repetitive, time-consuming tasks which can lead to errors. Thus, they will be free to fully exercise their expertise, at a microscope, a computer, or with a butterfly net in hand.

10.5. References

Camargo, A. and Sites, J. (2013). Species delimitation: A decade after the renaissance. In *The Species Problem – Ongoing Issues*, Pavlinov, I. (ed.). InTech, Rijeka.

Costello, M.J., May, R.M., Stork, N.E. (2013a). Can we name Earth's species before they go extinct? *Science*, 339(6118), 413–416.

Darwin, C.R. (1859). *On the Origin of Species by Means of Natural Selection, or the Preservation of Favoured Races in the Struggle for Life*, 1st edition. John Murray, London.

Dayrat, B. (2005). Toward integrative taxonomy. *Biological Journal of the Linnean Society*, 85, 407–415.

DeSalle, R., Egan, M.G., Mark, S. (2005). The unholy trinity: Taxonomy, species delimitation and DNA barcoding. *Philosophical Transactions of the Royal Society B*, 360, 1905–1916. doi: 10.1098/ rstb.2005.1722.

Dobzhansky, T. (1937). *Genetics and the Origin of Species*. Columbia University Press, New York.

Ducasse, J., Ung, V., Lecointre, G., Miralles, A. (2020). LIMES, a tool for comparing species partition. *Bioinformatics*, 36(7), 2282–2283.

Fouquet, A., Gilles, A., Vences, M., Marty, C., Blanc, M., Gemmell, N.J. (2007). Underestimation of species richness in neotropical frogs revealed by mtDNA analyses. *PLoS ONE*, 2(10), e1109.

Garnett, S.T. and Christidis, L. (2017). Taxonomy anarchy hampers conservation. *Nature*, 546(7656), 25–27.

Gewin, V. (2002). All living things, online. *Nature*, 418, 362–363.

Hebert, P.D.N., Cywinska, A., Ball, S.L., deWaard, J.R. (2003). Biological identifications through DNA barcodes. *Proceedings of the Royal Society of London*, 270(1512), 313–321. doi: 10.1098/rspb.2002.2218.

Hennig, W. (1950). *Grundzüge einer Theorie der phylogenetischen Systematik*. Deutscher Zentralverlag, Berlin.

Isaac, N.J.B., Mallet, J., Mace, G.M. (2004). Taxonomic inflation: Its influence on macroecology and conservation. *Trends in Ecology and Evolution*, 19(9), 464–469.

Kapli, P., Lutteropp, S., Zhang, J., Kobert, K., Pavlidis, P., Stamatakis, A., Flouri, T. (2016). Multi-rate Poisson tree processes for single-locus species delimitation under maximum likelihood and Markov chain Monte Carlo. *Bioinformatics*, 33, 1630–1638.

Larsen, B.B., Miller, E.C., Rhodes, M.K., Wiens, J.J. (2017). Inordinate fondness multiplied and redistributed: The number of species on Earth and the new pie of life. *Quaterly Review of Biology*, 92, 229–265.

Linnæus, C. (1758). *Systema naturæ per regna tria naturæ, secundum classes, ordines, genera, species, cum characteribus, differentiis, synonymis, locis, tome 1*. Typis Ioannis Thomae, Vienna.

Locey, K.J. and Lennon, J.T. (2016). Scaling laws predict global microbial diversity. *Proceedings of the National Academy of Sciences USA*, 113(21), 5970–5975.

Mallet, J. and Willmott, K. (2003). Taxonomy: Renaissance or Tower of Babel? *Trends in Ecology and Evolution*, 18, 57–59.

May, R.R. and Harvey, P.H. (2009). Species uncertainties. *Science*, 323(5915), 687.

Mayr, E. (1942). *Systematics and the Origin of Species, from the Viewpoint of a Zoologist*. Columbia University Press, New York.

Mayr, E. and Ashlock, P.D. (1991). *Principles of Systematic Zoology*, 2nd edition. McGraw-Hill, New York.

Meiri, S. and Mace, G.M. (2007). New taxonomy and the origin of species. *PLoS Biology*, 5(e194), 1385–1386.

Melville, J., Chaplin, K., Hipsley, C.A., Sarre, D.S., Sumner, J., Hutchinson, M. (2019). Integrating phylogeography and high-resolution X-ray CT reveals five new cryptic species and multiple hybrid zones among Australian earless dragons. *Royal Society Open Science*, 6(191166).

Miller, S.E. (2007). DNA barcoding and the renaissance of taxonomy. *Proceedings of the National Academy of Sciences USA*, 104(12), 4775–4776.

Miralles, A. (2021). *L'alpha-taxonomie au XXIe siècle : inventorier le Vivant à l'ère du numérique et de la 6e extinction.* Muséum national d'Histoire naturelle, Paris [Online]. Available at: https://hal.science/tel-03249035.

Miralles, A., Bruy, T., Wolcott, K., Scherz, M.D., Begerow, D., Beszteri, B., Bonkowski, M., Felden, J., Gemeinholzer, B., Glaw, F. et al. (2020). Repositories for taxonomic data: Where we are and what is missing? *Systematic Biology*, 69(6), 1231–1253.

Miralles, A., Puillandre, N., Vences, M. (2024). DNA barcoding in species delimitation: From genetic distances to integrative taxonomy. *Methods Mol. Biol.*, 2744, 77–104. doi: 10.1007/978-1-0716-3581-0_4.

Monaghan, M.T., Wild, R., Elliot, M., Fujisawa, T., Balke, M., Inward, D.J.G., Lees, D.C., Ranaivosolo, R., Eggleton, P., Barraclough, T.G. et al. (2009). Accelerated species inventory on madagascar using coalescent-based models of species delineation. *Systematic Biology*, 58(3), 298–311.

Padial, J.M. and De la Riva, I. (2006). Taxonomic inflation and the stability of species lists: The perils of ostrich's behavior. *Systematic Biology*, 55, 859–867.

Padial, J.M., Miralles, A., De la Riva, I., Vences, M. (2010). The integrative future of taxonomy. *Frontiers in Zoology*, 7(16).

Pante, E., Schoelinck, C., Puillandre, N. (2015) From integrative taxonomy to species description: One step beyond. *Syst. Biol.*, 64, 152–160. doi: 10.1093/sysbio/syu083.

Pons, J., Barraclough, T.G., Gomez-Zurita, J., Cardoso, A., Duran, D.P., Hazell, S., Kamoun, S., Sumlin, W.D., Vogler, A.P. (2006). Sequence-based species delimitation for the DNA taxonomy of undescribed insects. *Systematic Biology*, 55, 595–609.

Puillandre, N., Lambert, A., Brouillet, S., Achaz, G. (2012). ABGD, Automatic Barcode Gap Discovery for primary species delimitation. *Molecular Ecology*, 21, 1864–1877.

Puillandre, N., Brouillet, S., Achaz, G. (2020). ASAP: Assemble species by automatic partitioning. *Molecular Ecology Resources*, 21(2).

de Queiroz, K. (1998). The general lineage concept of species, species criteria, and the process of speciation. In *Endless Forms: Species and Speciation*, Howard, D.J. and Berlocher, S.H. (eds). Oxford University Press, New York.

de Queiroz, K. (2007). Species concepts and species delimitation. *Systematic Biology*, 56, 879–886.

Rissler, L.J. and Apodaca, J.J. (2007). Adding more ecology into species delimitation: Ecological niche models and phylogeography help define cryptic species in the black salamander (*Aneides flavipunctatus*). *Systematic Biology*, 56, 924–942.

Samadi, S. and Barberousse, A. (2006). The tree, the network, and the species. *Biological Journal of the Linnean Society*, 89(3), 509–521.

Seifert, B. (2020). The gene and gene expression (GAGE) species concept: An universal approach for all eukaryotic organisms. *Systematic Biology*, 69(5), 1033–1038.

Sites Jr., J.W. and Marshall, J.C. (2003). Delimiting species: A Renaissance issue in systematic biology. *Trends in Ecology and Evolution*, 18, 462–420.

Sites Jr., J.W. and Marshall, J.C. (2004). Operational criteria for delimiting species. *Annual Review of Ecolology, Evolution and Systematics*, 35, 199–227.

Tancoigne, E., Bole, C., Sigogneau, A., Dubois, A. (2011). Insights from Zootaxa on potential trends in zoological taxonomic activity. *Frontiers in Zoology*, 8(5).

Vences, M., Miralles, A., Brouillet, S., Ducasse, J., Fedosov, A., Kharchev, V., Kostadinov, I., Kumari, S., Patmanidis, S., Scherz, M.D. et al. (2011). iTaxoTools 0.1: Kickstarting a specimen-based software toolkit for taxonomists. *Megataxa*, 6, 77–92.

Vieites, D.R., Wollenberg, K.C., Andreone, F., Köhler, J., Glaw, F., Vences, M. (2009). Vast underestimation of Madagascar's biodiversity evidenced by an integrative amphibian inventory. *Proceedings of the National Academy of Sciences USA*, 106(20), 8267–8272.

Vogler, A.P. and Monaghan, M.T. (2006). Recent advances in DNA taxonomy. *Journal of Zoological Systematics and Evolutionary Research*, 45, 1–10.

Wheeler, Q.D. (2004). Taxonomic triage and the poverty of phylogeny. *Philosophical Transactions of the Royal Society B*, 359, 571–583.

Wheeler, Q.D., Raven, P.H., Wilson, E.O. (2004). Taxonomy: Impediment or expedient? *Science*, 303(5656), 285. doi: 10.1126/science.303.5656.285.

Wiens, J.J. (2007). Species delimitation: New approaches for discovering diversity. *Systematic Biology*, 56, 875–878.

Wiens, J.J. and Servedio, M.R. (2000). Species delimitation in systematics: Inferring diagnostic differences between species. *Proceedings of the National Academy of Sciences USA*, 267, 631–636.

Will, K.W., Mishler, B.D., Wheeler, Q.D. (2005). The perils of DNA barcoding and the need for integrative taxonomy. *Systematic Biology*, 54, 844–851.

Wilson, E.O. (2004). Taxonomy as a fundamental discipline. *Philosophical Transactions of the Royal Society B*, 359(1444), 739.

Yang, Z. and Rannala, B. (2010). Bayesian species delimitation using multilocus sequence data. *Proceedings of the National Academy of Sciences USA*, 107, 9264–9269.

11

Thiomargarita magnifica: A Giant from Marine Mangroves, Pushing the Limits of Bacteriology

11.1. Introduction

11.1.1. *A diverse bacterial world*

Microbial biodiversity is crucial in the living world. Prokaryotic microbes, which include bacteria and archeae, are particularly important in numerous biogeochemical cycles, and exhibit a wide variety of physiological and metabolic processes. These microorganisms also occupy numerous ecological niches: both aerobic and anaerobic. Thus, they are abundant in all environments. They are found in water (saltwater or freshwater), in soil and in marine sediments, and are even suspended in the air. Most of these organisms are single-celled (composed of a single cell, surrounded by a plasma membrane and a cell wall rich in peptidoglycan) and microscopic (only visible under an optical microscope). They do not have a nucleus to contain their genome, or membrane organelles characteristic of the eukaryotic cells making up plants, fungi and animals. Though generally, they are no bigger than 2 μm on average in diameter and/or length, there are some exceptionally large bacteria.

To date, giant bacteria have been found in marine environments, which are rich in organic matter. This is probably due to their energy requirements,

Chapter written by Olivier GROS, Silvina GONZALEZ-RIZZO, Nathalie ELISABETH and Jean-Marie VOLLAND.

owing to the fact of being giant within this microbial world. The majority of free giant bacteria so far described are autotrophic organisms capable of capturing carbon dioxide (CO_2) using light energy (photosynthetic bacteria) or the chemical energy released by sulfide oxidation (thioautotrophic chemosynthetic bacteria). The only currently known heterotrophic large bacteria are observed in the microbiome of eukaryotic organisms as symbiotes (Angert et al. 1993). Thus, there are numerous descriptions of these giant bacteria, but it seems clear that bacterial gigantism has evolved multiple times, as species of giant bacteria have been reported in at least seven branches.

11.1.2. *A microbial world is not always microscopic*

Numerous microscopic bacteria form large, multi-cellular structures, as happens with cyanobacteria (Komárek and Johansen 2015), cable bacteria (Pfeffer et al. 2012) and even magnetotactic bacteria (Keim et al. 2004). However, these bacteria are not considered to be true giants since their large structures result from the combination of thousands of cells, each individually microscopic. The concept of "giant bacteria" refers to bacterial cells which may or may not form larger, multicellular structures, but with the diameter or length of each individual cell far surpassing that of "conventional bacteria", generally measuring at least 10 µm in at least one dimension (Bizic et al. 2023). Over recent decades, multiple uncultivated bacteria have been announced as being the largest known prokaryotes. These include the bacterium Epulopiscium fishelsoni, which lives in the intestines of surgeonfish, and can be up to 80 µm in diameter and 600 µm in length (Clements and Bullivant 1991; Angert et al. 1993), certain bacteria of the genus *Beggiatoa* spp. (Larkin and Henk 1996; Jean et al. 2015), and the bacterium *Thiomargarita namibiensis* which can reach a maximum diameter of 750 µm, averaging 180 µm (Schulz et al. 1999).

11.1.3. *Physiological limitations linked to gigantism in bacteria*

Being of an exceptional size imposes major constraints, relating primarily to the distribution of metabolites throughout so vast a cell volume. For their nutritional needs, prokaryotes depend on passive diffusion, both for transfer of substances from the environment into the cell, and for intracellular transport. The larger the cell, the smaller its surface-to-volume ratio. This ratio can decrease by a factor of 100 in the case of a cell that is 10 times wider than normal (Gallet et al. 2017). Consequently, the time taken for solutes to diffuse through the cell wall and into the cytoplasm will be considerably longer, increasing from a few milliseconds in a conventional bacterium to several hours, or even several days, in the case of extremely large cells (Schulz and Jørgensen 2001). It is therefore vital for very large bacteria to find somewhere to remain during this exceedingly long diffusion time which, in theory, is prohibitive to the development of such cells.

Supposing that a nutrient supply can be secured for these prokaryotic giants, it is also crucial to generate a sufficient number of macromolecules (structural and functional proteins, RNA, etc.) to maintain harmonious cell function, on a physiological and metabolic level. Scott et al. (2010) showed that the larger a cell is, the greater the quantity of macromolecules it will need to have available. Subsequent work has shown that this significant production of proteins, essential to the development of giant bacteria, is linked to a higher number of ribosomes, rather than an increase in transcriptional or translational activity within the cells (Kempes et al. 2016).

Given the fairly high number of giant bacteria having now been described, it seems evident that these prokaryotic species have been able to develop innovative solutions to deal with these theoretical problems. One of the solutions adopted is to increase the surface-to-volume ratio by altering the shape of bacterial cells (Harris and Theriot 2018) and/or developing numerous folds in the cytoplasmic membrane (Angert 2006; Schulz 2006). Certain giant bacteria have opted to limit their metabolic activity to an area measuring only a few micrometers across, in contact with the cytoplasmic membrane, despite their size. This is made possible by the presence of a huge central vacuole, occupying 70–80% of the cell's volume, thereby reducing the active part of the cell (Volland et al. 2022).

Finally, having a very large volume of cytoplasm also poses the problem of accessibility to the genome for gene transcription and the rapid transport of mRNA throughout the cell. An original solution, and one which is rather unusual for prokaryotes, is to have multiple chromosomes (and thus, multiple genomes) distributed throughout the cell. The presence of more than ten chromosomes per cell is known as polyploidy. It is more common in bacteria and archeans than was previously thought (Oliverio and Katz 2014). All giant bacteria tested to date are polyploid, containing a large number of chromosomes: from a few hundred (Mendell et al. 2008; Ionescu et al. 2017) to hundreds of thousands in the same cell in the case of *Thiomargarita magnifica* (Volland et al. 2022).

Thus, certain giant bacteria have properties which are usually the preserve of eukaryotes, such as polyploidy (Ionescu et al. 2017; Angert 2021), and very recently, the presence of organelles enclosed by a membrane, containing DNA and ribosomes (Volland et al. 2022). The simplistic dichotomy between a prokaryote cell and a eukaryote cell, therefore, seems to be less rigid than we may think. The description of new, very large bacterial species is therefore essential in order to refine our knowledge and better understand the limits of what a bacterium can do and/or represent.

11.2. *Thiomargarita magnifica*

In 2022, a marine bacterium was described that was exceptional in all points. It lives attached to sunken organic substrates (the leaves and branches of the mangrove trees that cover the seabed) and inorganic ones too (oyster shells, plastic bags, old ropes, etc.) deposited on shallow marine sediments of mangroves (see Figure 11.1a). It appears in the form of long white filaments, around the size of a human eyelash, with a slightly flared base and an end part resembling a string of rosary beads (see Figure 11.1b).

Figure 11.1. *The giant bacterium* Thiomargarita magnifica *in its coastal marine environment. For a color version of this figure, see www.iste.co.uk/nicolas/biodiversity.zip*

COMMENT ON FIGURE 11.1.– *a) Underwater shot of mangrove sediments covered in plant debris, where a rich variety of microfauna lives (both prokaryotes and eukaryotes); b) a clump of* Thiomargarita magnifica *filaments attached to a submerged leaf of a red mangrove tree (*Rhizophora mangle*). Each filament, which is white due to the presence of intracellular elementary sulfur, terminates in a progressive narrowing, giving the appearance of a string of rosary beads, which is characteristic of this species.*

This giant bacterium exhibits three essential points, which are new in bacteriology: its size, its genome and its level of cellular complexity.

11.2.1. *A giant bacterium*

Thiomargarita magnifica cells are 10 mm in length on average, but may reach up to 20 mm (2 cm). Thus, the largest such specimens are up to 20,000 times the size of a typical bacterium such as *Escherichia coli*. To draw a parallel with humans, it is as if we had discovered a population of primates who were, on average, twice as tall as the world's tallest mountain, Mount Everest. This discovery is a truly stunning one in the world of microbiology.

The giant belongs to a group of bacteria which have been known for decades: the genus *Thiomargarita*, belonging to the branch Pseudomonadota, and the class Gamma-proteobacteria. Given its current position in the tree of life, this species is not the missing link between prokaryotes and eukaryotes. Thus, it is a bacterium – we can say that with 100% certainty – but it is one which is exceptional in every respect. It pushes the boundaries and expands the field of possibilities in bacteriology. However, size is not everything… it is merely a talking point.

11.2.2. *A bacterium with an exceptional genome*

This bacterium has a genome which is just as astonishing as its size. It is four times the size of the genome of a conventional bacterium, and thus codes for four times more proteins: a total of 11,800 proteins, which is on par with certain fungi. It is typically accepted that the majority of bacteria have only a single copy of their genome per cell, and that single copy floats freely in the cytoplasm. Thus, in the main, bacteria are haploid, though certain instances of oligoploidy have previously been described in the literature, with up to several tens of copies of the genome per cell in the case of certain species, such as bacteria of the genus *Achromatium* (Ionescu et al. 2017). However, as *Thiomargarita magnifica* does nothing the same as the rest, it also has the peculiarity of having a very great number of copies of its genome in each filament of *Thiomargarita magnifica*. Single cells have been found to contain up to 700,000 copies of the genome. *Thiomargarita magnifica* goes way beyond what was previously thought possible for a bacterium. By way of comparison, we humans have only two copies of our genome per cell. Certainly, our genome is bigger, but in total, a single *Thiomargarita magnifica* cell contains more DNA than one of our own cells. This bacterium is quite simply incredible!

11.2.3. *A bacterium with high cellular complexity*

Only animal or vegetable organisms, whether single- or multi-cellular, were thought to have a cell nucleus, and thus the possibility to compartmentalize their DNA. *Thiomargarita magnifica* has a level of cellular complexity that is hitherto totally unheard of for bacteria. The filaments of this species have the capacity to compartmentalize their numerous copies of their DNA in membrane-bound structures called "pepins" (Volland et al. 2022). These structures, containing DNA and ribosomes, are also the site of protein synthesis, confirming the presence of active ribosomes (Volland et al. 2022) and even enzyme activity linked to energy production, such as membrane ATP synthases. It could be, then, that pepins are structures similar to organelles: a term which is currently reserved for eukaryotic cells. Again, we see *Thiomargarita magnifica* pushing the boundaries of what bacteria are known to be able to do, and thus allowing us to redefine the outlines of the bacterial world.

11.2.4. *A giant colonizing the sediments in mangrove swamps in Guadeloupe*

Thiomargarita magnifica was discovered in marine mangroves in Guadeloupe. Mangroves are tropical ecosystems characterized by an intertidal forest straddling the boundary between the land and the sea. Thus, these ecosystems are home to enormous biodiversity, because they contain species from both the marine and land environments, but also species which are found nowhere else (Clough 2013). According to Spalding et al. (1997), the ebb and flow of the tides could favor the implantation of species of vegetation capable of adapting to these environments, with high levels of salinity (35g/L). The marine mangroves in the Caribbean are characterized by a low tidal range, meaning that the soils and sediments are never out of the water, and there is a great deal of accumulation of organic materials in the marine sediments. In Guadeloupe, the average tidal range is around thirty centimeters, which is relatively small when compared to certain mainland mangroves such as those in French Guyana or Asia, which can experience tidal ranges of up to 3.5 meters (Spalding et al. 1997). This low tidal range means that the salinity gradient of the environment is relatively stable.

The organic material, coming mainly from mangrove trees (leaves and branches) is decomposed anaerobically by sulfate-reducing bacteria, which produce large quantities of sulfides (H_2S, HS^-) (see Figure 11.2). Indeed, measurements taken in these environments revealed sulfide concentrations upwards of 1mM (Laurent et al. 2009). To compare, the "black smokers" of deep-ocean hydrothermal vents eject sulfides at a concentration of 100 μm on average. These sulfides are toxic for the majority of animals, as they block the last enzyme in the respiratory chain (Dorman et al. 2002).

Sulfides diffusing from the anoxic sediment are either chemically oxidized into sulfates in the water column, or used by aerobic bacteria situated at the oxic/anoxic interface, which biologically oxidize these sulfides into sulfates whilst obtained the energy they need for development (see Figure 11.2). These "sulfur-oxidizing" bacteria are said to be autotrophic as they use this chemical energy to capture CO_2 dissolved in the water and produce organic molecules. This process is known as chemosynthesis. The genomic and ultrastructural data obtained by Volland et al. (2022), and its presence at the oxic/anoxic interface in sulfide-rich mangrove sediments, show that *Thiomargarita magnifica* is a sulfur-oxidizing chemosynthetic bacterium.

Chemosynthesis is a process whereby organic compounds are synthesized from CO_2 dissolved in air or water. This process, which is comparable to photosynthesis by green plants, uses the energy released by the oxidation of inorganic chemical compounds (instead of light) to produce ATP or NADPH. This chemical energy can then be used to inject the carbon from the CO_2 into the Calvin–Benson cycle to produce organic molecules (sugars and amino acids) (Tunnicliffe et al. 2003). This process is performed by autotrophic microorganisms (bacteria or archeae), and generally requires consumption of oxygen, unlike photosynthetic organisms, which produce oxygen.

In marine mangroves, these sulfur-oxidizing chemosynthetic bacteria, belonging to different genera, may either be present freely in the sediments (*Achromatium*) or cover the surface of the sediments, forming veils (*Sulfuricurvum, Sulfurimonas, Thiovulum*) or mats composed of filamentary bacteria (*Beggiatoaceae*). These bacteria thus serve as primary producers, and assist the development of animal species capable of ingesting them and benefiting from the nutrients.

In the mangroves, some of these bacteria may also live in symbiotic partnerships with all sorts of marine invertebrates, ranging from protists (*Dysteria* spp., *Pseudovorticella* spp., *Zoothamnium* spp. (see Laurent et al. 2013; Grimonprez et al. 2018)) to more complex metazoans such as nematodes (Stilbonematidae (Himmel et al. 2009)), lucinid bivalves (*Phacoides pectinatus* (Frenkiel et al. 1996)) and jellyfish (*Cladonema* spp. (Abouna et al. 2015)).

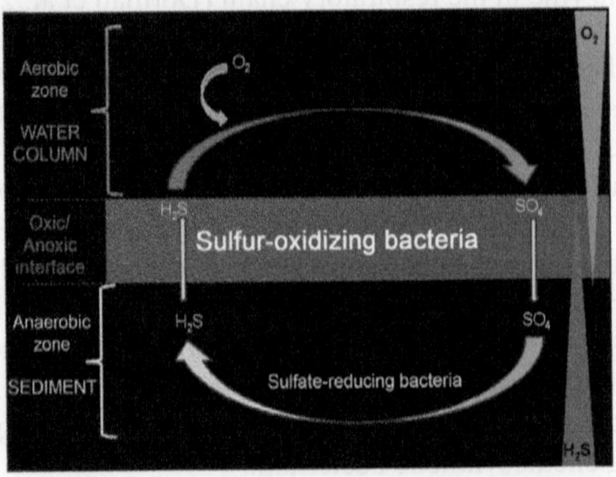

Figure 11.2. *Overview of the sulfur cycle in mangrove sediments. For a color version of this figure, see www.iste.co.uk/nicolas/biodiversity.zip*

COMMENT ON FIGURE 11.2.– *Sulfur-oxidizing bacteria are located at the oxic-anoxic interface, allowing them simultaneous access to sulfides from the anaerobic sediments and to oxygen from the aerobic water column. The gradients of sulfides and oxygen are shown on the right of this figure.*

11.3. Conclusion

It is currently estimated that over 99% of bacterial species in the environment have never yet been isolated, and thus, never cultured in a laboratory. Thus, the vast majority of prokaryotic biodiversity is known thanks to molecular analysis of environmental samples, which cannot provide information about the size of the microorganisms found. It is therefore likely that studies on bacterial biodiversity are partially biased in the way in which cells are collected for analysis. For example, aquatic bacterial communities are often filtered to exclude larger eukaryotic organisms, and therefore, giant bacteria. Overall, it is probable that there are other giant bacteria that have yet to be discovered. This demonstrates the need for observation in situ, in the field, to find abnormally long structures for microorganisms, like the filaments of *Thiomargarita magnifica*.

With respect to *Thiomargarita magnifica*, the next steps will be to discover whether this bacterium is spread throughout the Caribbean, whether it exists elsewhere in the world, or in other marine ecosystems. One of the short-term objectives will also be to determine its lifespan in this giant form, and to find out whether it is only present in the mangrove swamps at certain times of year.

Another important point is that we need to determine whether *Thiomargarita magnifica* is a unique case in evolutionary terms, or whether its peculiarity in terms of gigantism, multiple copies of genomes and compartmentalization of DNA is a more common phenomenon in giant bacteria, though they have not yet been found in many other organisms. The most recent results available seem to indicate that pepin-type structures do exist in other large bacteria present in the mangroves of Guadeloupe, confirming that *Thiomargarita magnifica* is merely the most visible representation of a more widespread phenomenon, but as of yet not well studied in giant bacteria.

The study of giant bacteria living as free-living organisms or in symbiotic relationships in different habitats or hosts is of considerable interest to scientists. These peaceful giants, if they could one day be cultivated, could also represent a study model that is unique in microbiology, allowing us to better understand the bacteria's adaptative capabilities. This is true at molecular level (DNA structure, management of polyploidy, cell division, etc.) and at metabolomic level (production of metabolites secondary to antimicrobial activity, production of specific proteins, etc.). Such advances would not be possible without a multidisciplinary approach, combining direct observations by researchers in the field with the development and application of new molecular tools in the laboratory.

To conclude, in the context of climate change and the upsets that come with it, mangroves will undoubtedly play an important role in coastal protection, but also as a haven for biodiversity. However, this ecosystem, which will be so crucial in the coming decades, is under threat from unrestrained urban development. In less than 50 years, it is estimated that around 25% of the world's mangroves will have disappeared. Insular mangroves and the biodiversity to which they are home are under threat, and we urgently need to make concerted efforts to study the microbial biodiversity found there, and continue to move science forward by studying these tropical ecosystems, which hold many wonders yet unknown.

11.4. References

Abouna, S., Gonzalez-Rizzo, S., Grimonprez, A., Gros, O. (2015). First description of bacterial sulfur-oxidizing symbiosis in a Cnidarian (Medusozoa) living in sulphidic shallow-water environments. *PLoS One*, 10(5), e0127625.

Angert, E.R. (2006). The enigmatic cytoarchitecture of *Epulopiscium* spp. In *Complex Intracellular Structures in Prokaryotes*, Shively, J.M. (ed.). Springer, Berlin/Heidelberg, 285–301.

Angert, E.R. (2021). Challenges faced by highly polyploid bacteria with limits on DNA inheritance. *Genome Biol. Evol.*, 13.

Angert, E.R., Clements, K., Pace, N. (1993). The largest bacterium. *Nature*, 362, 239–241.

Bizic, M., Brad, T., Ionescu, D., Barbu-Tudoran, L., Zoccarato, L., Aerts, J., Contarini, P.-E., Gros, O., Volland, J.-M., Popa, R. et al. (2023). Cave *Thiovulum* (*Candidatus* Thiovulum stygium) differs metabolically and genomically from marine species. *ISME J.*, 17, 340–353. doi: 10.1093/gbe/evab037.

Clements, K.D. and Bullivant, S. (1991). An unusual symbiont from the gut of surgeonfishes may be the largest known prokaryote. *J. Bact.*, 173, 5359–5362.

Clough, B. (2013). *Continuing the Journey Amongst Mangroves*. ISME, Okinawa.

Dorman, D.C., Moulin, F.J., McManus, B.E., Mahle, K.C., James, R.A., Struve, M.F. (2002). Cytochrome oxidase inhibition induced by acute hydrogen sulfide inhalation: Correlation with tissue sulfide concentrations in the rat brain, liver, lung, and nasal epithelium. *Toxicol. Sci.*, 65, 18–25.

Frenkiel, L., Gros, O., Mouëza, M. (1996). Gill ultrastructure in *Lucina pectinata* (Bivalvia: Lucinidae) with reference to hemoglobin in bivalves with symbiotic sulphur-oxidizing bacteria. *Mar. Biol.*, 125, 511–524.

Gallet, R., Violle, C., Fromin, N., Jabbour-Zahab, R., Enquist, B.J., Lenormand, T. (2017). The evolution of bacterial cell size: The internal diffusion-constraint hypothesis. *ISME J.*, 11, 1559–1568.

Grimonprez, A., Laurent, M.C.Z., Molza, A., Mansot, J.-L., Gros, O. (2018). Thioautotrophic ectosymbiosis in *Pseudovorticella* sp., a peritrich ciliate species colonizing wood falls in marine mangrove. *Eur. J. Protistol.*, 62, 43–55.

Harris, L.K. and Theriot, J.A. (2018). Surface area to volume ratio: A natural variable for bacterial morphogenesis. *Trends Microbiol.*, 26, 815–832.

Himmel, D., Maurin, L.C., Gros, O., Mansot, J.-L. (2009). Raman microspectrometry sulphur detection and characterization on the marine ectosymbiotic nematode *Eubostrichus dianae* (Desmodoridae, Stilbonematidae). *Biol. Cell.*, 101, 43–54.

Ionescu, D., Bizic-Ionescu, M., De Maio, N., Cypionka, H., Grossart, H.-P. (2017). Community-like genome in single cells of the sulfur bacterium *Achromatium oxaliferum*. *Nature*, 8(1), 455.

Jean, M.N.R., Gonzalez-Rizzo, S., Gauffre-Autelin, P., Lengger, S.K., Schouten, S., Gros, O. (2015). Two new *Beggiatoa* species inhabiting marine mangrove sediments in the Caribbean. *PLoS One*, 10(2), e0117832.

Keim, C.N., Martins, J.L., Abreu, F., Rosado, A.S., de Barros, H.L., Borojevic, R., Lins, U., Farina, M. (2004). Multicellular life cycle of magnetotactic prokaryotes. *FEMS Microbiol. Lett.*, 240, 203–208.

Kempes, C.P., Wang, L., Amend, J.P., Doyle, J., Hoehler, T. (2016). Evolutionary tradeoffs in cellular composition across diverse bacteria. *ISME J.*, 10, 2145–2157.

Komárek, J. and Johansen, J.R. (2015). Filamentous cyanobacteria. In *Freshwater Algae of North America*, 2nd edition, Wehr, J.D., Sheath, R.G., Kociolek, J.P. (eds). Academic Press, Boston. doi: 10.1016/B978-0-12-385876-4.00004-9.

Larkin, J.M. and Henk, M.C. (1996). Filamentous sulfide-oxidizing bacteria at hydrocarbon seeps of the Gulf of Mexico. *Microsc. Res. Tech.*, 33, 23–31.

Laurent, M.C.Z., Gros, O., Gaill, F., Le Bris, N. (2009). Sunken woods as a new sulphidic environment for chemosymbiosis in mangrove swamp. *Mar. Environ. Res.*, 67, 83–88.

Laurent, M.C.Z., Le Bris, N., Gaill, F., Gros, O. (2013). Dynamics of sunken woods colonization in relation to surface sulfide enrichments. *Mar. Environ. Res.*, 85–96.

Mendell, J.E., Clements, K.D., Choat, J.H., Angert, E.R. (2008). Extreme polyploidy in a large bacterium. *Proc. Natl. Acad. Sci. USA*, 105, 6730–6734.

Oliverio, A.M. and Katz, L.A. (2014). The dynamic nature of genomes across the tree of life. *Genome Biol. Evol.*, 6, 482–488.

Pfeffer, C., Larsen, S., Song, J., Dong, M., Besenbacher, F., Meyer, R.L., Kjeldsen, K.U., Schreiber, L., Gorby, Y.A., El-Naggar, M.Y. et al. (2012). Filamentous bacteria transport electrons over centimetre distances. *Nature*, 491, 218–221.

Schulz, H.N. (2006). The genus *Thiomargarita*. In *The Prokaryotes*, Dworkin, M., Falkow, S., Rosenberg, E., Schleifer, K.-H., Stackebrandt, E. (eds). Springer, New York.

Schulz, H.N. and Jorgensen B.B. (2001). Big bacteria. *Annu. Rev. Microbiol.*, 55, 105–137.

Schulz, H.N., Brinkhoff, T., Ferdelman, T.G., Marine, M.H., Teske, A., Jorgensen, B.B. (1999). Dense populations of a giant sulfur bacterium in Namibian shelf sediments. *Science*, 284, 493–495.

Scott, M., Gunderson, C.W., Mateescu, E.M., Zhang, Z., Hwa, T. (2010). Interdependence of cell growth and gene expression: Origins and consequences. *Science*, 330, 1099–1102.

Spalding, M.D., Blasco, E., Field, C.D. (1997). *World Mangrove Atlas. The International Society for Mangrove Ecosystems*. Cambridge University Press, Cambridge.

Tunnicliffe, V., Juniper, S.K., Sibuet, M. (2003). Reducing environments of the deep-sea floor. In *Ecosystems of the World: The Deep-Sea*, Tyler, P.A. (ed.). Elsevier, Amsterdam.

Volland, J.-M., Gonzalez-Rizzo, S., Gros, O., Tyml, T., Schulz, F., Udwary, D., Goudeau, D., Nath, N., Malmstrom, R., Guidi-Rontani, C. et al. (2022). A centimeter-long bacterium with DNA contained in metabolically active, membrane-bound organelles. *Science*, 376(6600), 1453–1458.

12

New Species of Freshwater Fish in France: Reasons and Impacts for Management

12.1. Introduction

Whilst certain taxonomic groups are under-studied due to lack of specialists, or due to their small size of rarity, vertebrates – particularly those which are useful to humans – have received far more attention since the beginning of systematics. "Knowledge of fish, born of our habit of eating them, must have been one of humankind's earliest acquired forms of knowledge", as Cuvier and Valenciennes (1828-1849) put it. Similarly, whilst certain areas such as the geographic poles or the ocean depths are difficult to access, the exploration of watercourses in mainland France requires neither specialist material nor enormous investment. Thus, French waterways did not seem to be a likely place to discover new species of ichthyofauna. However, knowledge has indeed advanced in recent years, and systematicians are continuing to unearth new species.

Of France's freshwater ichthyofauna species, 51 were described in the 18th and 19th centuries, in accordance with the International Code of Zoological Nomenclature, on the basis of morphological and anatomic features. These discoveries were made by Carl Linnaeus (1758), then Cuvier

Chapter written by Gaël DENYS, Agnès DETTAÏ, Henri PERSAT, Nicolas POULET and Philippe KEITH.

and Valenciennes (1828–1849), and some ten further species by Bloch (1785–1797), Risso (1810–1826) and Vallot (1837). The second half of the 19th century and the first half of the 20th saw a dramatic increase in regional fauna inventories, with descriptions of new species. However, certain naturalists interpreted morphological variations between individuals or ecotypes (such as river trout vs. sea trout) to indicate the existence of difference species, driving up the number of recorded species, peaking at inventories of 80–98 species (for example, see Blanchard 1866). At the same time, numerous species that were non-native, either to France as a whole or to certain river basins, were introduced and propagated for the purposes of pleasure fishing, aquaculture, scientific curiosity or mere "embellishment" (Persat et al. 2020).

The exhaustive review carried out by Spillmann (1961) considerably reduced the number of species to 55, demoting certain taxa to the rank of subspecies (see Figure 12.1). It was on this basis that the various national inventories that followed in the 1990s and early 2000s were carried out, integrating newly integrated species such as topmouth gudgeon and asp, taking the number of species to 83 (Allardi and Keith 1991; Billard 1997; Keith and Allardi 2001).

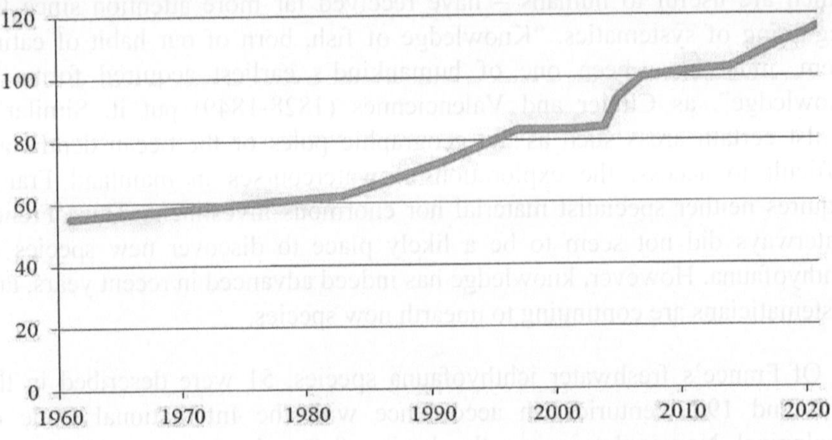

Figure 12.1. *Evolution of the number of freshwater fish species recorded in France over time, since Spillmann's (1961) review. For a color version of this figure, see www.iste.co.uk/nicolas/biodiversity.zip*

	Keith and Allardi (2001)	**Current taxonomic and geographic status**
Gudgeons	*Gobio gobio* (Linnaeus 1766)	*Gobio gobio* (Linnaeus 1758) → Rhine, Meuse, Seine, Loire and Rhône *Gobio occitaniae* Kottelat and Persat 2005 → Charente, Garonne, Languedoc-Roussillon, heads of the Loire and Rhône drainage basins, draining on the central plain *Gobio lozanoi* Doadrio and Madeira 2004 → Adour
Daces	*Leuciscus leuciscus* (Linaeus 1758)	*Leuciscus leuciscus* (Linnaeus 1758) → Rhine, Meuse, Seine and Rhône *Leuciscus bearnensis* (Blanchard 1866) → Adour *Leuciscus burdigalensis* Valenciennes 1844 → Loire, Charente, Garonne and Languedoc-Roussillon
Minnows	*Phoxinus phoxinus* (Linnaeus 1766)	*Phoxinus bigerri* Kottelat and Freyhof 2007 → Adour *Phoxinus csikii* Hankó 1922 → Rhine and Léman *Phoxinus dragarum* Denys, Dettai, Persat, Daszzkiewicz, Hautecoeur and Keith 2020) → Garonne *Phoxinus fayollarum* (Denys, Dettai, Persat, Daszzkiewicz, Hautecoeur and Keith 2020) → Loire *Phoxinus phoxinus* (Linnaeus 1758) → Rhine, Meuse, Seine *Phoxinus septimaniae* Kottelat and Freyhof 2007 → Rhône and Languedoc-Roussillon
Chubs	*Leuciscus cephalus* (Linnaeus 1766)	*Squalius cephalus* (Linnaeus 1758) → all river basins *Squalius laietanus* Doadrio, Kottelat and de Sostoa 2007 → Agly, Têt and Tech
Stone loaches	*Barbatula barbatula* (Linnaeus 1766)	*Barbatula hispanica* (Lelek 1987) → Adour *Barbatula leoparda* Gauliard, Dettai, Persat, Keith and Denys 2019 → Têt and Tech *Barbatula quignardi* (Băcescu-Mester 1967) → Garonne, coastal Mediterranean basin up to Lez *Barbatula spp* → Other river basins
Ninespined sticklebacks	*Pungitius pungitius* (Linnaeus 1758)	*Pungitius laevis* (Cuvier 1829–1849) → Seine, Loire *Pungitius pungitius* (Linnaeus 1758) → Meuse, Seine and Rhône *Pungitius vulgaris* (Mauduyt 1848) → Vienne basin to the Gironde estuary

Pikes	*Esox lucius* Linnaeus 1758	*Esox lucius* Linnaeus 1758 → All river basins *Esox aquitanicus* Denys, Dettai, Persat, Hautecoeur and Keith 2014 → From Charente to Adour
Graylings	*Thymallus thymallus* (Linnaeus, 1758)	*Thymallus ligericus* Persat, Weiss, Froufe, Secci-Petretto and Denys 2019 → Loire *Thymallus thymallus* (Linnaeus 1758) → Rhine, Meuse, Rhône
Sculpins	*Cottus gobio* Linnaeus, 1758 *Cottus petiti* Băcescu and Băcescu-Mester 1964	*Cottus aturi* Freyhof, Kottelat and Nolte 2005 → Adour *Cottus duranii* Freyhof, Kottelat and Nolte 2005 → Head of the Loire and Dordogne basins *Cottus gobio* Linnaeus 1758 → Rhine and Rhône *Cottus hispaniolensis* (Băcescu and Băcescu-Mester 1964) → Garonne *Cottus perifretum* Freyhof, Kottelat and Nolte 2005 → Seine and Loire *Cottus petiti* (Băcescu and Băcescu-Mester 1964) → Lez *Cottus rhenanus* Freyhof, Kottelat and Nolte 2005 → Meuse and Moselle *Cottus rondeleti* Freyhof, Kottelat and Nolte 2005 → Hérault

Table 12.1. *List of the systematic differentiations and taxonomic changes in the different genera since Keith and Allardi (2001), and areas of distribution according to Keith et al. (2020) and Denys et al. (2021). The names of the river basins are indicated after the arrows*

At that time, the inventory of native species present seemed to be complete (apart from a species described in 1964 – the Lez sculpin) and most species appeared to be common to neighboring countries. However, these certainties were overturned in the 2000s. The development of computing technology changed the way in which systematicians worked, allowing for the sharing, storage, comparison and complex analysis of large datasets, and the inclusion of pictures and videos. In parallel, the development of molecular systematics approaches based on the comparison of DNA sequences provided access to a new, less subjective type of data, which can be used to study similarities, the evolution of species and precisely identify samples, including for large-scale comparisons. Thus, Kottelat and Freyhof (2007), followed by Keith et al. (2011), took the list of species in France to over 100 – many of them specific (endemic) to certain river basins. The most recent inventory records 118 (Keith et al. 2020).

Thus, nine genera have been revised, with 26 additional species joining the taxonomic framework since the turn of the millennium (see Table 12.1). This list is by no means set in stone, because there are still groups of species that have not been reassessed, or at least, not completely. The taxonomic re-delimitation of species requires lengthy and complex work, including all data, and the verification of synonyms and past descriptions, which takes years.

In view of the lack of understanding from the public, managers and researchers that are not systematicians, who might consider these changes "taxonomic inflation", it is helpful to explain the sources of this diversity, the reasons for these changes, and their potential implications for the management of fish species.

12.2. Reasons for these changes

12.2.1. *Taking account of biogeographical context*

Humans, being terrestrial animals, naturally have a perception of their environment which is restricted by their own administrative boundaries. Our freshwater fish, being aquatic animals, are only geographically limited to their river basins. The hydrographic network in France contains some fifty-odd river basins. The main ones can be classified into three biogeographical provinces: the English Channel/the North Sea (the Rhine, Meuse and Seine), the Atlantic (the Loire, Garonne and Adour), and the Mediterranean (the Rhône and coastal basins). The network and its fauna have changed greatly over time. Three geological events have shaped the ichthyological composition since the Miocene (23 MYA) (Persat et al. 2020).

The Miocene was the period when the ichthyofauna represented by today's genera became established. This period was marked by Alpine folding, due to the slippage of the African plate beneath the Eurasian plate, which had consequences for the connections between the various basins. Freshwater blennies of the genus *Salariopsis* appear to have diverged from sea blennies during the late Miocene, between 8 and 5 MYA (for example, see Wagner et al. 2021). Other species from the Mediterranean Basin, such as the Catalan chub *Squalius laietanus*, the Rhone streber *Zingel asper*, and the Mediterranean barbel *Barbus meridionalis*, whose closest living relatives are species from the Balkans or the lower Danube, have hypothetical common ancestors, also estimated to have lived during that period (Perea et al. 2010; Stepien and Haponski 2015; Levin et al. 2019).

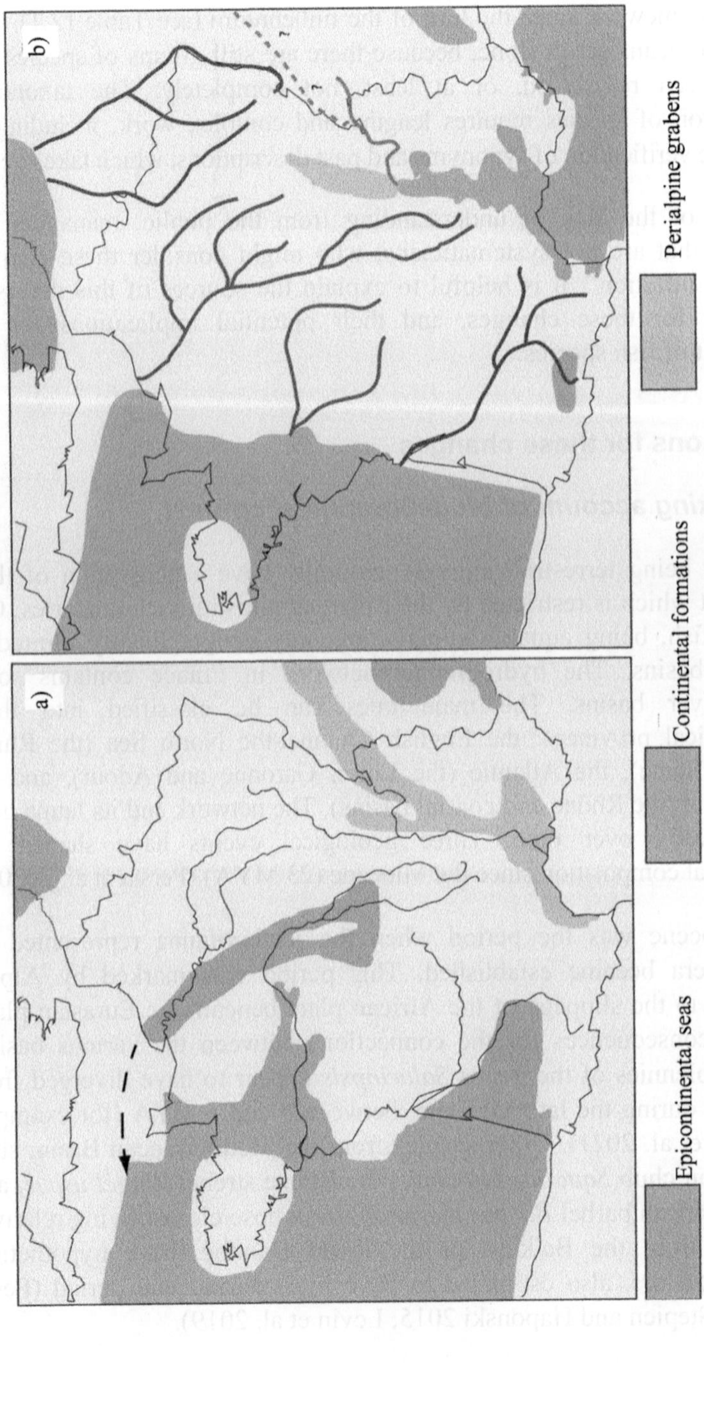

Figure 12.2. *France in the Miocene (a) and in the Pliocene (b); source: Persat et al. 2020. For a color version of this figure, see www.iste.co.uk/nicolas/biodiversity.zip*

Through these processes, the epicontinental seas began to be pushed back, allowing the creation of permanent rivers such as the Garonne, or a river that became the Seine and the Loire. The perialpine graben extended to form the Bresse Graben. On the Atlantic coast, Britanny became isolated, and the Rennes and Nantes basins and the plains of Aquitaine were still submerged (see Figure 12.2a). The North-Pyrenean Atlantic basins (which produced the Garonne and the Adour) seem to have become completely isolated from the Iberian catchment. On the Mediterranean side, the opening up of the Gulf of Lion and the eastward drift of the Corsica–Sardinia block caused a distension in the Pyrénées-Orientales department. This led to the formation of a series of meltwater basins (Cerdagne, Conflent, Roussillon), offering potential connections between the Sègre and Têt basins. In parallel, the perialpine graben closed up in the mid-Miocene (15 MYA). Thus, this marine basin was gradually replaced by a river basin.

The Messinian Crisis (the drying up of the Mediterranean by the closing of the Gibraltar Corridor, following the subduction of the African plate under the European plate, cutting off the connection between the Atlantic and the Mediterranean) would have favored connections between the drainage basins, notably along the Adriatic which, at the time, was merely a vast extension of the Pô Plain (Bianco 1990). However, recent works in geology suggest that there has always been some kind of connection between the two marine domains, and thus, the Mediterranean cannot have been completely dry (see Krijgsman et al. 2018)).

From the Pliocene (5.3–2.5 MYA), the raising of the Alps and the Jura, Vosges, Black Forest ensemble created, and then rearranged, multiple connections between the Danube, the Rhône in Valais and the Rhine in the Alps. The Loire joined the Seine, while Vienne, Indre and Sarthe together constituted an independent basin (see Figure 12.2b). The climate was temperate and the ichthyofauna seemed to be similar to the fauna we now see. The hydrographic network of the time is reflected in the distribution of the different species of ninespined sticklebacks, with *P. vulgaris* in the basins of Vienne, the Charente and the Gironde Estuary, while *P. laevis* is present in the Loire and Seine-Normandy Basins (Denys et al. 2018), with a divergence between 2.5 and 4.7 MYA (Wang et al. 2021). Allopatric speciation events began during this period, in the wake of the isolation of river basins, in graylings (*Thymallus* spp.), stone loaches (*Barbatula* spp.) and minnows (*Phoxinus* spp); however, we do not yet have details of the associated geological events (Sedivá et al. 2008; Corral-Lou et al. 2019; Weiss et al. 2020).

The Tertiary–Quaternary transition was characterized by a fall in global temperature, leading to several cyclical glacial episodes occurring from the start of the Pleistocene (1.8 MYA) up to the start of the Holocene, including four major ones over the past 500,000 years. All of France (indeed, all of central Europe) was then covered with tundra, and only the south of the Aquitaine Basin and Languedoc retained a sufficiently temperate climate for forests to survive, serving as a refuge for flora and fauna (Bertran et al. 2014; Vandenberghe et al. 2014).

During the glacial maxima, the Scandinavian Ice Sheet covered the Baltic countries and a large part of the United Kingdom. The Alpine Ice Sheet covered all the Alps and part of their foothills. This resulted in a major drop in sea level, allowing downstream connections between the Loire and the Vilaine, the Garonne and the Charente, and between the Catalan basins. The English Channel was also a potent fluvial axis, fed by at least the Oder, Elbe, Weser, Rhine, Meuse, Tamise and Seine (see Figure 12.3) (for example, see Lehmkuhl et al. 2021).

The different waves of cold considerably altered the flora and fauna (Hewitt 2004). Non-cryophilic species (excluding minnows, stone loaches and sculpins) north of the permafrost line died out, and were then replaced by new Dnieper/Danube ichthyofauna, which, taking advantage of the chaos during the interglacial periods, were able to colonize the affluents of the English Channel (Costedoat and Gilles 2009). The fauna in southwestern and southern France were in sheltered areas, and therefore survived the glaciations.

Iconic species such as the brown trout *Salmo trutta* and the threespined stickleback *Gasterosteus aculeatus* colonized the river basins of France during that period (Cortey et al. 2009; Fang et al. 2018).

Figure 12.3. *France during the last glacial maximum of the Pleistocene. Adapted from Lehmkuhl et al. 2021. Dark blue lines: rivers. Lilac lines: paleochannels on dry continental shelf. Grey lines: edge of dry continental shelf. Light green line: boundary of continuous permafrost. Light green dashed line: boundary of discontinuous permafrost. Dark green dashed line: northern boundary of forests. Light blue patches: extent of glaciers. For a color version of this figure, see www.iste.co.uk/nicolas/biodiversity.zip*

The current interglacial period, known as the Holocene (12 kYA to the present day) began with the end of the last Ice Age, the river basins became isolated from one another once more with the fall in sea level, leaving the hydrographic network as it currently stands. Rhine–Danube connections with the Meuse, Seine, Loire and Rhône basins were formed during the retreat of the Alpine glaciers.

Viewing the structuring of ichthyofauna in relation to biogeographic events, we see that there are three biogeographic domains:

– the Rhine–Danube domain (the Seine, Meuse and Rhine Basins), populated with species having essentially arisen during the Pleistocene glacial period, gaining access to colonize these basins *via* the English Channel;

– the Atlantic domain (the Loire, Garonne and Adour), comprising species which were already present before the Pleistocene glacial period;

– the Mediterranean domain (the Rhône and coastal Mediterranean basins), with a very high level of endemicity in Languedoc-Roussillon from species having originated in the late Miocene. The top of the Rhône basin includes species whose origin can be attributed to the diffluences with the Rhine and the Danube in the Pliocene, and to the natural colonizations during the Holocene via the Rhine.

12.2.2. *The concept of a species and the approach of integrative taxonomy*

The systematics of teleost fish presents particular problems in relation to other vertebrates, with polymorphisms (sometimes linked to the environment), physiological variations (temporary mating displays or amphihalinity), cases of genome duplications and polyploidy, or asexual reproduction and the phenomena of hybridization and introgression (the integration of genes from one species by another). For a long time, criteria based on "biological species" were employed – notably with respect to reproductive isolation (for example, see Mayr 1942). However, these criteria rely on potential interfertility in the natural environment, which is difficult to test, and is rarely tested. There are multiple cases of hybridizations between distant species, or even different genera, with viable and fertile descendants (Scribner et al. 2001). Thus, the biological concept best known to the public is not applicable in ichthyology, and reproductive isolation would be only one particular case (Turner 1999). From the time of Kottelat (1997), followed by Kottelat and Freyhof (2007), two other concepts have been applied: the phylogenetic concept (a species can be diagnosed by features that are unique to it – for example, see Cracraft 1983), and the evolutionary concept, which closely interlinks taxonomy, phylogeny, biogeography and evolutionary biology (Wiley and Mayden 2000). Now, whatever the concept used, the taxa described as subspecies are elevated to the rank of species (Kottelat 1997).

In the context of the recent taxonomic revisions carried out on French ichthyofauna, it is evolutionary criteria that are used, by integrative taxonomy methods, where the descriptions and delimitations of species are based on congruence between multiple independent types of data (data pertaining to genetics, morphology, reproduction, geographical distribution, ecology and behavior) (Padial et al. 2011). Thus, the taxonomic delimitations of all the species listed in Table 12.1 are supported by the combination of morphological, molecular and biogeographical criteria.

12.3. Impacts on management

In addition to its scientific importance, this taxonomic update has implications in terms of conservation (Isaac et al. 2004). Firstly, these changes have wrought confusion from a regulatory point of view. Indeed, these new species names are not on the list of species represented in the Decree of 17 December 1985, and thus cannot be returned to the water after capture if we strictly obey the regulation, although these are endemic species, which is absurd. Similarly, certain taxa are protected by the Decree of 8 December 1988, or that of 23 April 2008, protecting their spawning grounds (pike, sculpin) or listed in Appendix III of the Bern Convention (grayling). De facto, in the event of any change in scientific naming, it is sensible to give the newly named species the regulatory status enjoyed by the species listed originally, pending the possible revision of the list in question.

Next, given that we can only effectively protect that which we know, an undescribed species is a species in danger. However, unfortunately, there are very few studies carried out on the ecology of these new species, which poses a problem for their management and conservation (for example, see Dudgeon et al. 2006). They could have distinct ecological traits, which could skew biotic indices such as the *Indice Poisson Rivière* (IPR; Oberdorff et al. 2002). Incorrect identifications damage the state of knowledge in ecology and, consequently, hamper management efforts (Bortolus 2008). It is therefore crucial to raise awareness among all actors – the public, researchers and managers, and train them on how to identify the various species.

Finally, some of the newly described species have proved to be endemic or subendemic, which means that France is responsible for these species' survival. Comparing the major drainage basins (Rhine-Meuse, Seine-Normandie, Loire-Brittany-Charente, Adour-Garonne, Rhone and the Mediterranean basins), we can see that the southern basins which were more or less spared glaciation have lower biodiversity, but higher rates of endemicity than those in the north. They contain more introduced species than native ones (see Figure 12.4). Unfortunately, these introductions (be they fish or other aquatic organisms) represent threats for native species, which are particularly fragile because their distribution area is restricted: competition, predation, pathogen introduction, hybridization and introgression (for example, see Gozlan et al. 2010; Cucherousset and Olden 2011). There are many ways in which non-native species can be introduced and spread: the digging of channels for navigation and irrigation, aquaculture, biological pest control, "scientific curiosity" (Keith et al. 2020) and, in particular, pleasure fishing, which leads to the practice of "fish rearing" or "restocking" (Cucherousset et al. 2021). These practices involve the introduction of farmed individuals with the idea of buoying numbers in the exploited populations. The problem is that the individuals used for the repopulation often belong to species that are allochthonous to the target river or water plane. For example, in the southwest of France, restockings of gudgeons (*Gobio* ssp.), minnows (*Phoxinus* spp.) and pike (*E. lucius*) are common, and take place without giving consideration to the native species – in this case, *G. occitaniae*, *G. lozanoi* (Adour), *Phoxinus dragarum*, *P. bigerri* (Adour) and *E. aquitanicus*. Certain populations exhibit cases of hybridization, with individuals having alleles from multiple species, and thus a loss of the genetic integrity of the native or endemic species. This has repercussions on IUCN assessments, which examine the genetic status of populations (Garner et al. 2020); a hybrid population leads to the native species being considered extinct; this is what happened, notably, with *E. aquitanicus* (UICN comité français et al. 2019). Consequently, it is necessary, provided it is truly effective, for managers to prioritize the production of native lineages with a view to restocking – closed-circuit management at the scale of the original basins (Keith et al. 2020), and in particular, the protection of watercourses in which human activity has not yet influenced the ichthyological population.

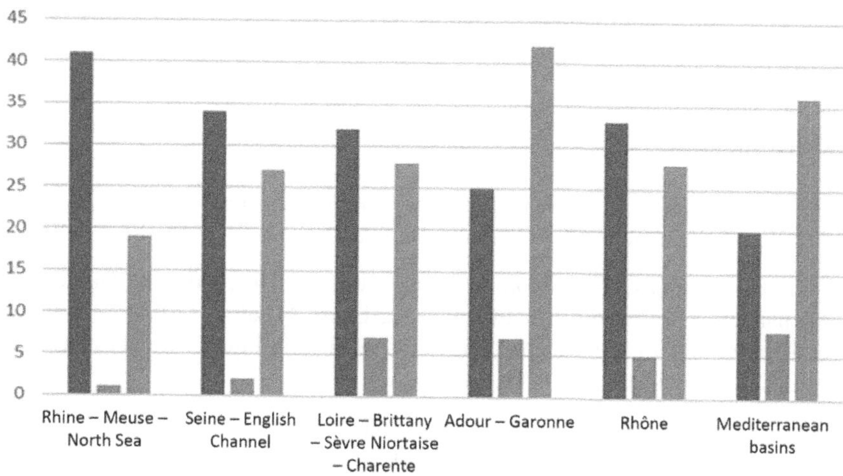

Figure 12.4. *Histogram representing the number of freshwater fish species that are native (blue), endemic (orange) or introduced (gray) in France's main river basins, in relation to the current state of knowledge of species nativity, which may still be underestimated. For a color version of this figure, see www.iste.co.uk/nicolas/biodiversity.zip*

12.4. Conclusion

The newly described ichthyological diversity reflects the history and evolution of species in France, but the new species also need to be described on the basis of their biological characteristics (tolerance of temperature range, spawning season, resistance to species or parasites introduced or encouraged by environmental changes). These characteristics may have an important impact on these species' resilience in the event of dramatic environmental change, and on their usefulness in scientific contexts (as indicator species) or economic ones (aquaculture, fishing).

It is essential to take account of these newly described species in ecological studies and environment monitoring. For decades, our lack of knowledge of France's ichthyofauna has been a barrier to the establishment of appropriate measures to protect endemic species that are part of our natural heritage.

The establishment of a management policy for each basin is crucial, in order to preserve over 20 million years of local evolutionary history as far as possible, against the steamroller of climate change and the globalization of fauna and flora, characteristic of the crisis we are experiencing in the Anthropocene.

12.5. References

Allardi, J. and Keith, P. (1991). *Atlas préliminaire des poissons d'eau douce de France*. Muséum national d'Histoire naturelle, Paris.

Bertran, P., Andrieux, E., Antoine, P., Coutard, S., Deschodt, L., Gardère, P., Hernandez, M., Legentil, C., Lenoble, A., Liard, M. et al. (2014). Distribution and chronology of Pleistocene permafrost features in France: Database and first results. *Boreas*, 43(3), 699–711.

Bianco, P.G. (1990). Potential role of the palaeohistory of the Mediterranean and Paratethys basins on the early dispersal of Euro-Mediterranean freshwater fishes. *Ichthyological Exploration of Freshwaters*, 1(2), 167–184.

Billard, R. (1997). *Les poissons d'eau douce des rivières de France. Identification, inventaire et répartition des 83 espèces*. Delachaux & Niestlé, Lonay.

Blanchard, E. (1866). *Les poissons des eaux douces de la France : anatomie, physiologie, description des espèces, mœurs, instincts, industrie, commerce, ressources alimentaires, pisciculture, législation concernant la pêche*. J.-B. Baillière et fils, Paris.

Bloch, M.E. (1785–1797). *Ichtyologie ou histoire naturelle, générale et particulière des poissons : avec des figures enluminées, dessinées d'après nature*. François de la Garde, Berlin.

Bortolus, A. (2008). Error cascades in the biological sciences: The unwanted consequences of using bad taxonomy in ecology. *AMBIO: A Journal of the Human Environment*, 37(2), 114–118.

Corral-Lou, A., Perea, S., Aparicio, E., Doadrio, I. (2019). Phylogeography and species delineation of the genus *Phoxinus* Rafinesque, 1820 (Actinopterygii: Leuciscidae) in the Iberian Peninsula. *Journal of Zoological Systematics and Evolutionary Research*, 57(4), 926–941.

Cortey, M., Vera, M., Pla, C., García-Marín, J. (2009). Northern and Southern expansions of Atlantic brown trout (*Salmo trutta*) populations during the Pleistocene. *Biological Journal of the Linnean Society*, 97(4), 904–917.

Costedoat, C. and Gilles, A. (2009). Quaternary pattern of freshwater fishes in Europe: Comparative phylogeography and conservation perspective. *The Open Conservation Biology Journal*, 3(1), 36–48.

Cracraft, J. (1983). Species concepts and speciation analysis. In *Current Ornithology*, Johnston, R.F. (ed.). Plenum Press, New York, 159–187.

Cucherousset, J. and Olden, J.D. (2011). Ecological impacts of non-native freshwater fishes. *Fisheries*, 36(5), 215–230.

Cucherousset, J., Lassus, R., Riepe, C., Millet, P., Santoul, F., Arlinghaus, R., Buoro, M. (2021). Quantitative estimates of freshwater fish stocking practices by recreational angling clubs in France. *Fisheries Management and Ecology*, 28(4), 295–304.

Cuvier, G. and Valenciennes, A. (1828–1849). *Histoire naturelle des poissons*. F.G. Levrault, Paris.

Denys, G.P.J., Persat, H., Dettai, A., Geiger, M.F., Freyhof, J., Fesquet, J., Keith, P. (2018). Genetic and morphological discrimination of three species of ninespined stickleback *Pungitius* spp. (Teleostei, Gasterosteidae) in France with the revalidation of *Pungitius vulgaris* (Mauduyt, 1848). *Journal of Zoological Systematics and Evolutionary Research*, 56, 77–101.

Denys, G.P.J., Perea, S., Dettai, A., Gauliard, C., Persat, H., Doadrio, I. (2021). Revalidation of the Spanish stone loach *Barbatula hispanica* (Lelek 1987) (Teleostei, Nemacheilidae) according to morphological and mitochondrial data. *International Journal of Limnology*, 57(10), 11.

Dudgeon, D., Arthington, A., Gessner, M., Kawabata, Z., Knowler, D., Lévêque, C., Naiman, R., Prieur-richard, A., Soto, D., Stiassny, M. et al. (2005). Freshwater biodiversity: Importance, threats, status and conservation challenges. *Biological Reviews*, 81(2), 163–182.

Fang, B., Merilä, J., Ribeiro, F., Alexandre, C., Momigliano, P. (2018). Worldwide phylogeny of three-spined sticklebacks. *Molecular Phylogenetics and Evolution*, 127, 613–625.

Garner, B., Hoban, S., Luikart, G. (2020). IUCN red list and the value of integrating genetics. *Conservation Genetics*, 21(5), 795–801.

Gozlan, R., Britton, J., Cowx, I., Copp, G. (2010). Current knowledge on non-native freshwater fish introductions. *Journal of Fish Biology*, 76(4), 751–786.

Hewitt, G. (2000). The genetic legacy of the Quaternary ice ages. *Nature*, 405, 907–913.

Isaac, N., Mallet, J., Mace, G.M. (2004). Taxonomic inflation: Its influence on macroecology and conservation. *Trends in Ecology & Evolution*, 19(9), 464–469.

Keith, P. and Allardi, J. (2001). *Atlas des poissons d'eau douce de France*. Muséum national d'Histoire naturelle, Paris.

Keith, P., Persat, H., Feuteun, E., Allardi, J. (2011). *Les poissons d'eau douce de France*. Muséum national d'Histoire naturelle, Paris.

Keith, P., Poulet, N., Denys, G., Changeux, T., Persat, H. (2020). *Les poissons d'eau douce de France*. Muséum national d'Histoire naturelle, Paris.

Kottelat, M. (1997). European freshwater fishes. An heuristic checklist of the freshwater fishes in Europe (exclusive of former USSR), with an introduction for non-systematists and comments on nomenclature and conservation. *Biologia Bratislava*, 52(5), 1–271.

Kottelat, M. and Freyhof, J. (2007). *Handbook of European Freshwater Fishes*, Publications Kottelat, Cornol.

Krijgsman, W., Capella, W., Simon, D., Hilgen, F., Kouwenhoven, T., Meijer, P., Sierro, F., Tulbure, M., van den Berg, B., van der Schee, M. et al. (2018). The Gibraltar corridor: Watergate of the Messinian salinity crisis. *Marine Geology*, 403, 238–246.

Lehmkuhl, F., Nett, J.J., Pötter, S., Schulte, P., Sprafke, T., Jary, Z., Antoine, P., Wacha, L., Wolf, D, Zerboni, A. et al. (2021). Loess landscapes of Europe – Mapping, geomorphology, and zonal differentiation. *Earth-Science Reviews*, 215, 103496.

Levin, B.A., Gandlin, A.A., Simonov, E.S., Levina, M.A., Barmintseva, A.E., Japoshvili, B., Mugue, N.S., Mumladze, L., Mustafayev, N.J., Pashkov, A.N. et al. (2019). Phylogeny, phylogeography and hybridization of Caucasian barbels of the genus *Barbus* (Actinopterygii, Cyprinidae). *Molecular Phylogenetics and Evolution*, 135, 31–44.

Linnaeus, C. (1758). *Systema naturae per regna tria naturae, secundum classes, ordines, genera, species, cum characteribus, differentii, synonymis, locis*. Impensis Direct, Holmiae.

Mayr, E. (1942). *Systematics and the Origin of Species from the Viewpoint of a Zoologist*. Columbia University Press, New York.

Oberdorff, T., Pont, D., Hugueny, B., Belliard, J., Berrebi, R., Porcher, J.P. (2002). Adaptation et validation d'un indice poisson (FBI) pour l'évaluation de la qualité biologique des cours d'eau français. *Bulletin Français de la Pêche et de la Pisciculture*, 365–433.

Padial, J., Miralles, A., De la Riva, I., Vences, M. (2010). The integrative future of taxonomy. *Frontiers in Zoology*, 7(1), 16.

Perea, S., Böhme, M., Zupančič, P., Freyhof, J., Šanda, R., Özuluğ, M., Abdoli, A., Doadrio, I. (2010). Phylogenetic relationships and biogeographical patterns in Circum-Mediterranean subfamily Leuciscinae (Teleostei, Cyprinidae) inferred from both mitochondrial and nuclear data. *BMC Evolutionary Biology*, 10(1), 265.

Persat, H., Keith, P., Denys, G. (2020). Biogéographie et historique de la mise en place des peuplements ichtyologiques de France métropolitaine. In *Les poissons d'eau douce de France*, Keith, P., Poulet, N., Denys, G., Changeux, T., Persat, H. (eds). Muséum national d'Histoire naturelle, Paris, 43–97.

Risso, A. (1810). *Ichthyologie de Nice, ou, histoire naturelle des poissons du département des Alpes Maritimes*. F. Schoell, Paris.

Risso, A. (1826). *Histoire naturelle des principales productions de l'Europe méridionale et particulièrement de celles des environs de Nice et des Alpes Maritimes*. F.-G. Levrault, Paris.

Scribner, K., Page, K., Bartron, M. (2000). Hybridization in freshwater fishes: A review of case studies and cytonuclear methods of biological inference. *Reviews in Fish Biology and Fisheries*, 10(3), 293–323.

Šedivá, A., Janko, K., Šlechtová, V., Kotlík, P., Simonović, P., Delic, A., Vassilev, M. (2008). Around or across the Carpathians: Colonization model of the Danube basin inferred from genetic diversification of stone loach (*Barbatula barbatula*) populations. *Molecular Ecology*, 17(5), 1277–1292.

Spillmann, C.J. (1961). *Faune de France. Poissons d'eau douce de France.* Editions Paul Lechevalier, Paris.

Stepien, C.A. and Haponski, A.E. (2015). Taxonomy, distribution, and evolution of the Percidae. In *Biology and Culture of Percid Fishes*, Kestemont, P., Dabrowski, K., Summerfelt, R. (eds). Springer, Dordrecht.

Turner, G.F. (1999). What is a fish species? *Reviews in Fish Biology and Fisheries*, 9, 281–297.

UICN comité français, MNHN, SFI, AFB (2019). *La Liste rouge des espèces menacées en France*. UICN, Gland.

Vallot, J.N. (1837). *Ichthyologie française, ou, histoire naturelle des poissons d'eau douce de la France*. Imprimerie E. Frantin, Dijon.

Vandenberghe, J., French, H., Gorbunov, A., Marchenko, S., Velichko, A., Jin, H., Cui, Z., Zhang, T., Wan, X. (2014). The last permafrost maximum (LPM) map of the northern hemisphere: Permafrost extent and mean annual air temperatures, 25–17 ka BP. *Boreas*, 43(3), 652–666.

Wagner, M., Zogaris, S., Berrebi, P., Freyhof, J., Koblmüller, S., Magnan, P., Laporte, M. (2021). Diversity and biogeography of Mediterranean freshwater blennies (Blenniidae, *Salaria*). *Diversity and Distributions*, 27(9), 1832–1847.

Wang, Y., Wang, Y., Zhao, Y., Kravchenko, A., Merilä, J., Guo, B. (2021). Phylogenomics of northeast asian *Pungitius* sticklebacks. *Diversity and Distributions*, 28(12), 2610–2621.

Weiss, S., Gonçalves, D., Secci-Petretto, G., Englmaier, G., Gomes-Dos-Santos, A., Denys, G.P.J., Persat, H., Antonov, A., Hahn, C., Taylor, E. et al. (2020). Global systematic diversity, range distributions, conservation and taxonomic assessments of graylings (Teleostei: Salmonidae; *Thymallus* spp.). *Organisms Diversity & Evolution*, 21, 25–42.

Wiley, E.O. and Mayden, R.L. (2000). The evolutionary species concept. In *Species Concepts and Phylogenetic Theory – A Debate*, Wheeler, Q.D. and Meier, R. (eds). Columbia University Press, New York, 70–89.

13

Effects of Sampling Bias in Estimating Phylodiversity in the Southern Ocean

To know that we know what we know, and that we do not know what we do not know, this is true knowledge.

Henry David Thoreau, Walden, 1854

The biological diversity of the living world, in the form of genomes, species and ecosystems, is threatened by numerous anthropic factors (Diaz and Malhi 2022). In order to effectively manage the biodiversity crisis, we need to take precise measurements of its variation, from a local to a global scale, which requires as comprehensive a knowledge as possible of the biological identity, spatiotemporal distribution and ecology of species. A thorough knowledge of the state of biodiversity is also crucial in order to preserve the Southern Ocean, one of the areas of the world which is being altered the most rapidly by global warming (Griffiths et al. 2017; Rogers et al. 2020).

13.1. Why study the Southern Ocean and its biodiversity?

The importance of the Southern Ocean in the global ecosystems' functioning cannot be ignored. It is unique in its connection to other oceans, and the site of a great deal of primary production. It plays a major role in global oceanic currents, which carry heat and atmospheric CO_2 to the depths, and thus participates in global climate regulation (Griffiths et al. 2017; Rogers et al. 2020). The Southern Ocean is characterized by specific

Chapter written by Anna KONDRATYEVA.

environments (low and stable temperatures, seasonal variations of sea ice extent and light regime (Griffiths et al. 2009)), which are home to unique and often endemic flora and fauna (Clarke and Johnston 2003), adapted to these extreme conditions (De Broyer et al. 2014; Peck 2018). Nearly 10,000 marine species have been described, to date, and the total number is estimated to be between 17,000 and 20,000, owing to the heterogeneity of the environment (De Broyer et al. 2014).

Numerous recent scientific studies indicate changes in the Antarctic ecosystem's functioning and in the structure and composition of the communities of species living there (Morley et al. 2020; Rogers et al. 2020; Brasier et al. 2021). In the majority of cases, these are consequences of human activity, whether indirect (due to climate change, in particular, or the arrival of invasive species and pollution) or direct (due to living resource exploitation (fishing) and tourism). These pressures are being exerted unequally, depending on the latitudes within the Southern Ocean (Morley et al. 2020). Global warming, for example, is particularly pronounced around the Antarctic Peninsula, and the subantarctic islands (Constable et al. 2014).

Therefore, stepping up research around the Southern Ocean is more important than ever. It is absolutely crucial that we estimate how the diversity of marine species in this ocean, which are habituated to extreme but stable conditions for nearly 34 million years (Crame 2018), will respond to global changes. For these reasons, keeping the Southern Ocean in a good state of health ought to be a major priority for managers and political decision-makers (Rogers et al. 2020).

13.2. Knowledge of marine biodiversity in the Southern Ocean

> In order to promote international cooperation in scientific investigation in Antarctica […], scientific observations and results from Antarctica shall be exchanged and made freely available. (Antarctic Treaty, article 3.1.c)

Our knowledge of Southern Ocean – one of the most difficult regions of the world's oceans to access – began with James Cook's exploratory expeditions in 1772–1775, and has been enriched by numerous scientific missions from the mid-19th Century onwards, such as the studies on benthos and plankton carried out aboard the vessels *Discovery* and *HMS Challenger*.

Since then, the number of species newly discovered and described has been ever increasing. However, despite recent efforts in Antarctic research, there are still numerous remote areas that are difficult to access, which pose particular challenges for the collection of biodiversity data, along the shores of the Antarctic mainland and in the open sea, particularly at great depths (Brandt et al. 2014).

Studies of small groups of species at local or regional scales in the Southern Ocean have increased in number, and initiatives have been undertaken to pool the data that these studies have used and produced. For example, the *Census of Antarctic Marine Life* (CAML) (Schiaparelli et al. 2013), the largest ever exploration of marine life in the Southern Ocean, collected a vast quantity of biotic and abiotic data (data on the distribution and abundance of benthic and pelagic species) and integrated them with preexisting historical collections from the *SCAR-MarBIN* programs[1]. One of the aims was to create the world's largest portal of georeferenced data on Antarctic marine biodiversity with an associated taxonomic framework[2]. Today, this database is continuously evolving, and yields the necessary information to assess changes in the diversity and/or the structure of communities under growing pressure from climate change, fishing and tourism (Van de Putte et al. 2021).

13.3. Sampling bias in data on Antarctic marine biodiversity

The Antarctic Marine Biodiversity Database is one of the regional nodes of the global data network pertaining to the occurrences of marine species, widely used in biogeography, macroecology and conservation (García-Roselló et al. 2015). These large public databases are the Global Biodiversity Information Facility (GBIF)[3] and Ocean Biodiversity Information System (OBIS)[4].

The data fed into these repositories come from a range of sources: scientific studies, inventories for national and/or regional atlases, museum collections and citizen science programs. Often, these data are collected without a specific sampling protocol, or with a protocol tailored to objectives

1 SCAR-Marine Biodiversity Information Network (www.scarmarbin.be).

2 See: www.data.biodiversity.aq.

3 See: www.gbif.org.

4 See: www.obis.org.

other than those of previous analyses (for example, suited for taxonomy, but not for distribution). Consequently, these data often suffer from complex sampling bias (Meyer et al. 2015), which could influence our understanding of the biodiversity and reduce the effectiveness of conservation measures driven by those data (Garnett and Christidis 2017).

The types of shortfall affecting these databases are due to incomplete knowledge concerning biodiversity. The primary lacks of knowledge that can be found in the data on species relate to their taxonomy (Linnean shortfall) (see Hortal et al. 2015 for references to the various lacks of knowledge on biodiversity), their distribution (Wallacean shortfall), abundance (Prestonian shortfall) and their evolutionary path and phylogenetic relations (Darwinian shortfall). Hortal et al. (2015) redefined a lack of knowledge on species' response to environmental or climatic conditions (Hutchinsonian shortfall) and proposed two others: one linked to limited knowledge of species' traits (Raunkiaeran shortfall) and another linked to limited knowledge of biotic interactions (Eltonian shortfall). Finally, there is also temporal shortfall, which could relate to particular seasons or years (Newbold 2010; Courter et al. 2013).

It is unlikely that the shortfalls in biodiversity data will soon be resolved, even with the development of technologies for more automated observation and data collection (underwater video and photography, environmental DNA). It is therefore essential for scientists who study biological diversity using diversity indices take account of the intrinsic shortfalls associated with the data they are using; in particular, to estimate the degree to which they may affect the analyses, results and conclusions.

13.4. Biodiversity measurements for the Southern Ocean

Species richness, calculated as the number of species found simultaneously in a given area, is the most commonly used measurement of diversity, including in the Southern Ocean (Griffith 2010; Stark et al. 2014; Friedlander et al. 2020). However, biodiversity cannot and must not be reduced to a simple list of species that populate a particular habitat, because it is far more than that. Indeed, as measurements of species richness do not take account of the biological identity of each species, they are treated as all being equivalent without taking account of their evolutionary history and phylogenetic relationship, which can be represented by a phylogenetic tree (the tree of life).

On the other hand, the use of molecular phylogenies makes it possible to take account of their evolutionary history, and the cumulative evolutionary differences may account for variations in characteristics (known or as yet unknown) and in different species' tolerance to climate change and other natural and anthropic disturbances. The species living in the Southern Ocean are isolated from those in the surrounding oceans by strong marine currents, and have been for the past 35–40 million years (Tripati et al. 2005). This may also be reflected in their phylogenies, endowing them with unique biological characteristics which merit protecting.

Finally, the phylogenetic approach enables us to unite numerous taxa through their evolutionary relationships in a single analysis, and to assess the extent of their evolutionary potential. Thus, species need to be evaluated as a function of their contribution to the tree of life (Faith 1992).

The most commonly used measure of phylodiversity is "phylogenetic diversity" (PD) (Faith 1992). PD calculates a value for a set of terminal taxa in a given geographic region as the sum of the lengths of branches as the shortest distance between the root of the phylogenetic tree and the taxa at the tips of the terminal branches. The predominance of this index is partly due to its simplicity, however, PD is strongly correlated with species richness, and its value increases with the addition of new species into the computation, which means that this index can be very strongly influenced by sampling biases.

The dependence of the PD index on species richness is, today, compensated by simple statistical standardization. For example, taking the residuals of a regression between PD and richness, instead of using the raw PD values. However, this does not address the problem of sampling bias. The data collected on CAML expeditions have been useful in elucidating the phylogenetic links within numerous taxonomic groups, thanks to modern molecular-biology techniques (Schiaparelli et al. 2013). These phylogenies can now be accessed for future studies of Antarctic phylogeography which are beginning to emerge.

13.5. Effects of sampling bias on the calculation of phylodiversity indices

Studies on the spatial distribution of phylodiversity require three types of data on the taxa being studied: their taxonomy, their georeferenced occurrences and their position on a phylogenetic tree. Each data type will lead to specific shortfall in the calculation of the phylodiversity indices. In this section, we shall briefly describe the three associated shortfalls – taxonomic, spatial and evolutionary – and put forward a few solutions.

13.5.1. *Taxonomic shortfall and phylodiversity*

Taxonomic shortfall is omnipresent in biodiversity research, and relates both to living and extinct species. This shortfall stems from the gap between the number of species formally described and the number of species which actually exist, and also the lack of universal rules governing taxonomy (Garnett and Christidis 2017; Diniz-Filho et al. 2023). The description of a new species within a clade – for example, following the split of one species into two – automatically increases the PD index (Robuchon et al. 2019). Failure to recognize new valid classifications in conservation programs would lead to the associated phylodiversity being underestimated.

Certain taxonomists draw on the phylogenetic concept of a species (Cracraft 1989), which holds that the two populations of individuals are classed as distinct species if they have a common ancestor but are morphologically and/or genetically different (Zachos 2016). Other taxonomists – for example, in ornithology – favor the concept of a biological species (Mayr 1963) whereby true species should not be able to produce fertile hybrids (Zachos 2016). Working on the hypothesis of "taxonomic inflation" (Isaac et al. 2004), the preference for the concept of a phylogenetic species would split the concept of a biological species, and thus increase the number of species recorded (Agapow et al. 2004).

In the Southern Ocean, certain groups of organisms have been much more widely studied than others, such as fish, birds and marine mammals, meaning that species of invertebrates which are far more numerous but also less well known to science have been understudied (Hughes et al. 2021). In addition, we see incorrect identifications of individuals observed and/or collected opportunistically or non-exhaustively, which also skews the value of the PD index observed. Thus, the view of the diversity of species in the Southern Ocean is still incomplete, except for certain regional studies, chiefly on the benthos, where the species are more accessible than pelagic species (Costello et al. 2017).

13.5.1.1. *Some short-term solutions*

Phylodiversity studies need to keep up the pace with changes in species classification in the wake of new descriptions and reviews of synonymous names (Costello et al. 2017), and come into line with taxonomic names of reference (e.g. the Register of Antarctic Species[5] for the Southern Ocean, which draws on the best possible taxonomic expertise (Van de Putte et al. 2021), the WoRMS[6] register, or France's taxonomic names referential TaxRef (Gargominy et al. 2022)). It is also important to choose a working unit for taxonomy which is suited to the study, rather than sticking to a single definition of what a species is (Kling et al. 2018). Another solution would be to use phylodiversity indices that are less sensitive to topological changes in the terminal branches of phylogenetic trees, instead being based on the average branch length more sensitive to deep branch length changes (Pavoine et al. 2017).

13.5.2. *Evolutionary shortfall and phylodiversity*

Evolutionary or Darwinian shortfall refers to a lack of knowledge of the tree of life and the evolution of lineages, species and their biological characteristics (Diniz-Filho et al. 2013). In the course of recent decades, the rapid progress in molecular genetics, bio-informatics and phylogenetic reconstruction incorporating fossil information have led to a rapid increase in the resolution and availability of time-calibrated phylogenies. However, there remains a degree of phylogenetic uncertainty linked to phylogenetic relations between species and the time-calibration of phylogenetic trees (using fossils or the associated geological strata), which results in uncertainty as to branch length, evolutionary timescales and tree topology.

The unit of a phylogenetic branch length may be based on the rate of change in molecular characteristics in the DNA sequences used: we then

---

5 RAS (ed.) (2023). *Register of Antarctic Species* (www.ras.biodiversity.aq).

6 WoRMS Editorial Board (2023). *World Register of Marine Species* (www.marinespecies.org).

refer to a phylogram (Cavender-Bares et al. 2004). Alternatively, it may be deduced from the age of a node in millions of years, using molecular clock models, calibrated in the presence of fossils: we then refer to a chronogram (Kling et al. 2018). Each of these theoretical structures (phylogenies) leads to different phylodiversity distribution patterns, reflecting distinct macro-evolutionary processes. Indeed, certain studies have found that, in certain cases, chronograms give PD values that are noticeably different to those calculated with phylograms for the same group of species in the same geographic zone (Elliott et al. 2018; Allen et al. 2019; Jantzen et al. 2019).

Revisions of phylogenetic tress may increase or decrease the number of existing species (following a split or the extinction of a species branch), which influences the value of global phylodiversity (Robuchon et al. 2019). This shortfall can therefore divert attention from species or groups of species which need to be a priority for conservation efforts.

13.5.2.1. *Some short-term solutions*

The uncertainties in calibration of phylogenies need to be taken into account when calculating phylodiversity. The choice of tree type and unit of length of the branches must be made on the basis of what we wish to target in our conservation efforts (diversification rate, divergence or survival over time, etc.).

If species for which we have available data on occurrence are not currently included in phylogenetic trees, it is possible to produce a tree to study these species using genetic sequences available on GenBank (though it is also time-consuming). Reconstructing a new phylogenetic tree for the group of species being studied could yield slightly different phylodiversity values compared to a phylogeny that is a subset tree cut from larger phylogeny (Jantzen et al. 2019).

Two quicker solutions would be: 1) to "manually" add missing species at the level of the node corresponding to their genus or family, using the free software R[7] (but for this, a known taxonomic classification is needed). For example, for ray-finned fish (*Actinopterygii*), there is an R package specifically dedicated to this issue (FishPhyloMaker) (see Nakamura et al. 2021); or 2) to use synthesis trees from phylogenetic data hubs, bringing

7 R Core Team 2021.

together trees from various sources (the Open Tree of Life[8], Time Tree[9] or TreeBase Web[10]). Phylodiversity measures obtained on the basis of synthesis phylogenies and phylogenies specifically constructed for a targeted study have been found to be strongly correlated (Li et al. 2019).

13.5.3. *Spatial shortfall and phylodiversity*

Spatial shortfall is a consequence of a lack of Wallacean knowledge (named for Alfred Russel Wallace 1823–1913), and refers to insufficient knowledge of species' geographic distribution. Spatial shortfall is probably the type of shortfall that has received most attention from the community of ecologists and conservation biologists, though it is taxonomic shortfall which actually has most impact (Hortal et al. 2015).

In general, the sampling rate is highest in the most easily accessible places in space and time. The rate may even be influenced by a country's sociopolitical situation (Zizka et al. 2020). In the oceans, more sampling takes place in immediate proximity to the shore and to shipping lines, at depths of up to 1,000 meters (Griffiths et al. 2014). The sampling effort is highly unequal between regions of the world ocean where coastal records (that is, ones taken 5km or less from the shore) represent 30–50% of records for the majority of taxa, apart from a few exceptions in the best studied groups (Hughes et al. 2021). Thus, spatial shortfall has more of an impact when it comes to rare and/or endemic species, as they are found less frequently.

Spatial and taxonomic shortfalls are heavily dependent on the spatial scale at which the occurrences are recorded and analyzed. The higher the spatial resolution, the greater the amount of knowledge necessary to define the area of distribution of a species, and also the lack of knowledge. For example, at the scale of the whole Earth, we would have perfect knowledge of the distribution of all known species, and thus spatial shortfall would not exist. The spatial scale at which diversity is studied needs to be adapted to fit the study in question, but also to the ecology of the species.

---

8 See: www.tree.opentreeoflife.org/.

9 See: www.timetree.org/.

10 See: www.treebase.org.

Finally, while occurrence databases are increasingly well supplied, they suffer from various problems in terms of data quality, which may have an impact on their use (Moudrý and Devillers 2020). To cite just a few, they often include problems linked to the geographical coordinates and timecodes (input errors, lack of coordinates, inadequate resolution) and information duplication between the databases.

The distribution of phylodiversity observed is directly impacted by analyses of occurrence data with a spatial shortfall, which can produce erroneous results, which risk influencing decisions on conservation.

13.5.3.1. *Some short-term solutions*

In order to reduce spatial shortfall, modeling efforts have been made, incorporating environmental information (ecological niche of the species) and occurrence data to estimate the probability of occurrence of species (rather than true occurrences) on sites that have not been sampled. For example, the project AquaMaps[11] offers modeled spatial distribution maps for marine species, at a 0.5° scale. Using species distribution models (SDMs), it is possible to fill in certain gaps in Wallacean knowledge, but only if the raw data have previously undergone automated cleansing (Moudrý and Devillers 2020; Zizka et al. 2020). SDMs based on biased data will always be biased, but it is possible to improve them. For example, by integrating the accessibility index into the distribution models (Monsarrat et al. 2019; Dubos et al. 2022).

It is also possible to integrate phylogenetic information directly into lineage distribution models, by modeling the probability of occurrence of each terminal branch (and thus, the species, which is tantamount to using conventional SDMs) and of the internal branches (Pollock et al. 2015), thus partly circumventing the taxonomic and spatial shortfalls.

Finally, a method developed by Chao for estimating biological diversity has been adapted to the PD index (Hsieh et al. 2016). This method uses rarefaction curves (interpolation) and prediction (extrapolation) curves which normalize the sample size at undersampled sites on the basis of the predefined number of individuals/species, and facilitate the comparison of estimated phylodiversity between zones with unequal sampling. However,

11 See: www.aquamaps.org.

this method is also suited to the handling of real abundance data (number of individuals), which are often unavailable for numerous marine species.

13.6. Conclusion

With a growing stock of data concerning biodiversity, it is important, today, to ensure that those data are representative, that they are properly stored, handled, processed and shared. We need further resources to collect data and carry out field studies that are standardized for the groups of organisms targeted. The new GBIF strategic framework 2023–2027 (Miller 2022) recognizes the main challenges that need to be addressed to improve the quality of raw data on biodiversity, and puts forward numerous solutions. To improve the quality of future data, it is desirable to use the norms and standards for information on biodiversity established by the Taxonomic Databases Working Group (TDWG)[12], which would enable us to develop, adapt and promote best practices across the user community. In order to prevent needless duplication of data transformation efforts, initiatives to share data-analysis methods are being established, such as France's *Pôle national des données sur la biodiversité* (PNDB, National Biodiversity Data Hub), whose aims are in keeping with the FAIR principle (that data should be findable, accessible, interoperable and reusable), and it is important to develop an infrastructure, tools and services for better understanding of biodiversity and the environment.

The scientific community continues to recommend conservation measures based on the existing empirical data, though those data are biased in relation to many taxonomic groups. The role of these types of studies is also to indicate these shortfalls and attempt to have them corrected, or at least taken into account, so that future studies can use them as effectively as possible. Existing biodiversity data must not paralyze conservation efforts, but should, on the contrary, encourage them. They should also encourage sampling campaigns in areas which have so far been insufficiently sampled. Free data sharing on biodiversity should be a priority for the scientific community, because open science is one of the crucial vectors for scientific knowledge production and diffusion.

12 See: www.tdwg.org.

13.7. References

Agapow, P.M., Bininda-Emonds, O.R., Crandall, K.A., Gittleman, J.L., Mace, G.M., Marshall, J.C., Purvis. A. (2004). The impact of species concept on biodiversity studies. *The Quarterly Review of Biology*, 79(2), 161–79.

Allen, J.M., Germain-Aubrey, C.C., Barve, N., Neubig, K.M., Majure, L.C., Laffan, S.W., Soltis, P.S. (2019). Spatial phylogenetics of Florida vascular plants: The effects of calibration and uncertainty on diversity estimates. *iScience*, 11, 57–70.

Brandt, A., Griffiths, H., Gutt, J., Linse, K., Schiaparelli, S., Ballerini, T., Danis, B., Pfannkuche, O. (2014). Challenges of deep-sea biodiversity assessments in the Southern Ocean. *Advances in Polar Science*, 25(3), 204–212.

Brasier, M.J., Barnes, D., Bax, N., Brandt, A., Christianson, A.B., Constable, A.J., Downey, R., Figuerola, B., Griffiths, H., Gutt, J. et al. (2021). Responses of Southern Ocean seafloor habitats and communities to global and local drivers of change. *Front. Mar. Sci.*, 8, 622721.

Cavender-Bares, J., Ackerly, D.D., Baum, D.A., Bazzaz, F.A. (2004). Phylogenetic overdispersion in Floridian oak communities. *The American Naturalist*, 163(6), 823–843.

Clarke, A. and Johnston, N.M. (2003). Antarctic marine benthic diversity. *Oceanography and Marine Biology. An Annual Review*, 41, 55–57.

Constable, A.J., Melbourne-Thomas, J., Conrey, S.P., Arrigo, K.R., Barbraud, C., Barnes, D.K., Bindoff, N.L., Boyd, P.W., Brandt, A., Costa, D.P. et al. (2014). Climate change and southern ocean ecosystems I: How changes in physical habitats directly affect marine biota. *Glob. Change Biol.*, 20, 3004–3015.

Costello, M.J. Basher, Z., McLeod, L., Asaad, I., Claus, S., Vandepitte, L., Yasuhara, M., Gislason, H., Edwards, M., Appeltans, W. et al. (2017). Methods for the study of marine biodiversity. In *The GEO Handbook on Biodiversity Observation Networks*, Walters, M. and Scholes, R.J. (eds). Springer, Berlin.

Courter, J.R., Johnson, R.J., Stuyck, C.M., Lang, B.A., Kaiser, E.W. (2013). Weekend bias in citizen science data reporting: Implications for phenology studies. *International Journal of Biometeorology*, 57(5), 715–720.

Cracraft, J. (1989). Speciation and its ontology: The empirical consequences of alternative species concepts for understanding patterns and processes of differentiation. In *Speciation and its Consequences*, Otte, D. and Endle, J.A. (eds). Sinauer Associates Inc., Sunderland.

Crame, J.A. (2018). Key stages in the evolution of the Antarctic marine fauna. *Journal of Biogeography*, 45(5), 986–994.

De Broyer, C., Koubbi, P., Griffiths, H.J., Raymond, B., d'Udekem d'Acoz, C. (2014). *Biogeographic Atlas of the Southern Ocean*. Scientific Committee on Antarctic Research, Cambridge.

Diaz, S. and Malhi, Y. (2022). Biodiversity: Concepts, patterns, trends, and perspectives. *Annual Review of Environment and Resources*, 47, 31–63.

Diniz-Filho, J.A.F., Loyola, R.D., Raia, P., Mooers, A.O., Bini, L.M. (2013). Darwinian shortfalls in biodiversity conservation. *Trends in Ecology & Evolution*, 28, 689–695.

Diniz-Filho, J.A.F., Jardim, L., Guedes, J.J., Meyer, L., Stropp, J., Frateles, L.E. Pinto, R.B., Lohmann, L.G., Tessarolo, G., de Carvalho, C.J.B. et al. (2023). Macroecological links between the Linnean, Wallacean, and Darwinian shortfalls. *Frontiers of Biogeography*, 15(2), e59566.

Dubos, N., Préau, C., Lenormand, M., Papuga, G., Monsarrat, S., Denelle, P., Le Louarn, M., Heremans, S., May, R., Roche, P. et al. (2022). Assessing the effect of sample bias correction in species distribution models. *Ecological Indicators*, 145(109487).

Elliott, M.J., Knerr, N.J., Schmidt-Lebuhn, A.N. (2018). Choice between phylogram and chronogram can have a dramatic impact on the location of phylogenetic diversity hotspots. *Journal of Biogeography*, 45, 2190–2201.

Faith, D.P. (1992). Conservation evaluation and phylogenetic diversity. *Biological Conservation*, 61, 1–10.

Friedlander, A.M., Goodell, W., Salinas-De-León, P., Ballesteros, E., Berkenpas, E., Capurro, A.P., Cárdenas, C.A., Hüne, M., Lagger, C., Landaeta, M.F. et al. (2020). Spatial patterns of continental shelf faunal community structure along the Western Antarctic peninsula. *PLoS One*, 15, 1–19.

García-Roselló, E., Guisande, C., Manjarrés-Hernández, A., González-Dacosta, J., Heine, J., Pelayo-Villamil, P., González-Vilas, L., Vari, R.P., Vaamonde, A., Granado-Lorencio, C. et al. (2015). Macroecological patterns and GBIF data. *Global Ecology and Biogeography*, 24, 335–347.

Gargominy, O., Tercerie, S., Régnier, C., Ramage, T., Dupont, P., Daszkiewicz, P., Poncet, L. (2022). TAXREF, référentiel taxonomique pour la France : méthodologie, mise en œuvre et diffusion. Report, PatriNat (OFB-CNRS-MNHN), Muséum national d'Histoire naturelle, Paris.

Garnett, S. and Christidis, L. (2017). Taxonomy anarchy hampers conservation. *Nature*, 546, 25–27.

Griffiths, H.J. (2010). Antarctic marine biodiversity – What do we know about the distribution of life in the Southern Ocean? *PLoS One*, 5(8), e11683.

Griffiths, H.J., Barnes, D.A., Linse, K. (2009). Towards a generalized biogeography of the Southern Ocean benthos. *Journal of Biogeography*, 36(1), 162–177.

Griffiths, H.J., Van de Putte, A.P., Danis, B. (2014). Data distribution: Patterns and implications. In *Biogeographic Atlas of the Southern Ocean*, De Broyer, C., Koubbi, P., Griffiths, H.J., Raymond, B., d'Udekem d'Acoz, C. (eds). SCAR, Cambridge.

Griffiths, H.J., Meijers, A.J.S., Bracegirdle, T.J. (2017). More losers than winners in a century of future Southern Ocean seafloor warming. *Nature Climate Change. Nature*, 7(10), 749–754.

Hortal, J., de Bello, F., Diniz-Filho, J.A.F. Lewinsohn, T.M., Lobo, J.M., Ladle, R.J. (2015). Seven shortfalls that beset large-scale knowledge of biodiversity. *Annual Review of Ecology, Evolution, and Systematics*, 46, 523–549.

Hsieh, T.C., Ma, K.H., Chao, A. (2016). iNEXT: An R package for rarefaction and extrapolation of species diversity (Hill numbers). *Methods in Ecology and Evolution*, 7(12), 1451–1456.

Hughes, A.C., Orr, M.C., Ma, K., Costello, M.J., Waller, J., Provoost, P., Yang, Q., Zhu, C., Qiao, H. (2021). Sampling biases shape our view of the natural world. *Ecography*, 44, 1259–1269.

Isaac, N.J.B., Mallet, J., Mace, J.M. (2004). Taxonomic inflation: Its influence on macroecology and conservation. *Trends in Ecology & Evolution*, 19(9), 464–469.

Jantzen, J.R., Whitten, W.M., Neubig, K.M., Majure, L.C., Soltis, D.E., Soltis, P.S. (2019). Effects of taxon sampling and tree reconstruction methods on phylodiversity metrics. *Ecology and Evolution*, 9, 9479–9499.

Kling, M.M., Mishler, B.D., Thornhill, A.H., Baldwin, B.G., Ackerly, D.D. (2018). Facets of phylodiversity: Evolutionary diversification, divergence and survival as conservation targets. *Philosophical Transactions of the Royal Society B*, 374(1763), 20170397.

Li, D., Trotta, L., Marx, H.E., Allen, J.M., Sun, M., Soltis, D.E., Soltis, P.S., Guralnick, R.P., Baiser, B. (2019). For common community phylogenetic analyses, go ahead and use synthesis phylogenies. *Ecology*, 100, e02788.

Mayr, E. (1963). *Animal Species and Evolution*. Belknap Press Harvard University, Cambridge.

Meyer, C., Kreft, H., Guralnick, R., Jetz, W. (2015). Global priorities for an effective information basis of biodiversity distributions. *Nature Communications*, 6, 8221.

Miller, J. (2022). GBIF's (global biodiversity information facility) new strategic framework to address biodiversity data needs. *Biodiversity Information Science and Standards*, 6, e91093.

Monsarrat, S., Boshoff, A.F., Kerley, G.I.H. (2019). Accessibility maps as a tool to predict sampling bias in historical biodiversity occurrence records. *Ecography*, 42, 125–136.

Morley, S.A., Abele, D., Barnes, D.K.A., Cárdenas, C.A., Cotté, C., Gutt, J., Henley, S.F., Höfer, J., Hughes, K.A., Martin, S.M. et al. (2020). Global drivers on Southern Ocean ecosystems: Changing physical environments and anthropogenic pressures in an Earth system. *Front. Mar. Sci.*, 7, 547188.

Moudrý, V. and Devillers, R. (2020). Quality and usability challenges of global marine biodiversity databases: An example for marine mammal data. *Ecological Informatics*, 56, 101051.

Nakamura, G., Richter, A., Soares, B.E. (2021). FishPhyloMaker: An R package to generate phylogenies for ray-finned fishes. *Ecological Informatics*, 66, 101481.

Newbold, T. (2010). Applications and limitations of museum data for conservation and ecology, with particular attention to species distribution models. *Progress in Physical Geography*, 34(1), 3–22.

Pavoine, S., Bonsall, M.B., Dupaix, A., Jacob, U., Ricotta, C. (2017). From phylogenetic to functional originality: Guide through indices and new developments. *Ecological Indicators*, 82, 196–205.

Peck, L. (2018). Antarctic marine biodiversity: Adaptations, environments and responses to change. *Oceanography and Marine Biology: An Annual Review*, 56, 105–236.

Pollock, L.J., Rosauer, D.F., Thornhill, A.H., Kujala, H., Crisp, M.D., Miller, J.T., McCarthy, M.A. (2015). Phylogenetic diversity meets conservation policy: Small areas are key to preserving eucalypt lineages. *Philosophical Transactions of the Royal Society B*, 370(1662), 20140007.

Robuchon, M., Faith, D.P., Julliard, R., Leroy, B., Pellens, R., Robert, A., Thévenin, C., Véron, S., Pavoine, S. (2019). Species splitting increases estimates of evolutionary history at risk. *Biological Conservation*, 235, 27–35.

Rogers, A.D., Frinault, B.A.V., Barnes, D.K.A., Bindoff, N.L., Downie, R., Ducklow, H.W., Friedlaender, A.S., Hart, T., Hill, S.L., Hofmann, E.E. et al. (2020). Antarctic futures: An assessment of climate-driven changes in ecosystem structure, function, and service provisioning in the Southern Ocean. *Annu. Rev. Mar. Sci.*, 12, 87–120.

Schiaparelli, S., Danis, B., Wadley, V., Michael Stoddart, D. (2013). The census of antarctic marine life: The first available baseline for antarctic marine biodiversity. In *Adaptation and Evolution in Marine Environments*, Verde, C. and di Prisco, G. (eds). Springer, Berlin.

Sites, J.W. and Marshall, J.C. (2004). Operation criteria for delimiting species. *Annual Review of Ecology, Evolution, and Systematics*, 35, 199–227.

Stark, J.S., Kim, S.L., Oliver, J.S. (2014). Anthropogenic disturbance and biodiversity of marine benthic communities in Antarctica: A regional comparison. *PLoS One*, 9(e98802).

Tripati, A. and Elderfield, H. (2005). Deep-sea temperature and circulation changes at the Paleocene–Eocene thermal maximum. *Science*, 308, 1894–1898.

Van de Putte, A.P., Griffiths, H.J., Brooks, C., Bricher, P., Sweetlove, M., Halfter, S., Raymond, B. (2021). From data to marine ecosystem assessments of the Southern Ocean: Achievements, challenges, and lessons for the future. *Frontiers in Marine Science*, 8(637063).

Zachos, F.E. (2016). *Species Concepts in Biology*. Springer, Cham.

Zizka, A., Antunes Carvalho, F., Calvente, A., Rocio Baez-Lizarazo, M., Cabral, A., Coelho, J.F.R., Colli-Silva, M., Fantinati, M.R., Fernandes, M.F., Ferreira-Araújo, T. et al. (2020). No one-size-fits-all solution to clean GBIF. *PeerJ*, 8, e9916.

14

Standardization, Accessibility of Research Data and Open Science

> Innovative scientific research has a crucial role in addressing global challenges – ranging from healthcare and climate change to renewable energy and natural resources management. The speed and depth of this research depends on fostering collaboratives exchanges between different communities and assuring its widest dissemination. The exchange of ideas, knowledge and data emerging is fundamental for human progress and is part of the core of OECD values. Thus, I am very pleased that the OECD has taken the lead in developing principles and standards to facilitate access to research data generated with public funding. (OECD 2007, p. 4)

14.1. Why talk about standards, open data and open science?

Academic scientific research exists only because of the publication of the research results in academic journals, in all domains. However, practices differ noticeably from one research domain to another. Computer science has long since adopted the principles of open science, with the concept of free software, which emerged in the 1980 with Richard Stallman's project GNU (Gnu's Not Unix) and the creation of the first General Public License (GPL) in 1989, setting the legal rules for the publication of computer source codes. However, it was years before this practice began to spread to other

---

Chapter written by Visotheary UNG.

disciplines. The Berlin Declaration (2003) and the position taken by the OECD (2007) were turning points in the openness of research. Recent years have been conclusive for open science.

14.1.1. *Issues*

Open science is a growing movement. Its aim is to make scientific research more accessible, transparent and collaborative. It is based on the belief that research should be freely accessible to all, and that transparency and information exchange are crucial to move science forwards. In the context of open science, scientists are encouraged to publish their work in open-access scientific journals. In addition, they are encouraged to share their research data and source codes, to facilitate the verification and reproduction of scientific results, thus improving the quality of scientific research.

Open science encourages scientific collaboration, and synergizes skills in order to address complex questions. It is a way to democratize scientific research and ensuring as many people as possible can access it. In addition, it encourages citizen science, by inviting the general public to help collect data (for example, the projects INPN-Espèces (n.d.), Les Herbonautes (n.d.) and iNaturalist (n.d.)), or assist in data analysis.

The opening up of the world of research is a relatively new phenomenon (less than 20 years), if we discount computer science, which led the way by some distance. Change in practice would not come until the national and international context became conducive to it (see Figure 14.1).

Figure 14.1. *Illustration of the Turing Way project by Scriberia (this illustration is created by Scriberia with the Turing Way community. Used under a CC-BY 4.0 license. doi: 10.5281/zenodo.3332807). For a color version of this figure, see www.iste.co.uk/nicolas/biodiversity.zip*

14.1.2. *French national context*

In France, the law for a digital republic, known as the Lemaire Law, published on 16 October 2016 (Légifrance 2016) marked a turning point for the world of academia. Indeed, the first section relates to the circulation of data and knowledge. The other two relate to protection of individuals in the digital society, and access to digital technology for all. Notably, for the first time, it introduces openness by default for what are considered "*public data*":

> All data which are, or should (legally or voluntarily) be published or made available to the public, and which are produced or collected by a state, a regional authority, a parastatal organization, in the course of their public-service activities. (Wikipedia n.d.a)

The open science movement was then strengthened by the publication of an initial national plan for open science 2018–2021 (MESR 2018) and the creation of a Web portal, *Ouvrir la science*[1]. This site states that open science is:

> [...] the unrestricted dissemination of results, methods and products of scientific research. It is based on the opportunity presented by the digital transition to develop open access to publications and – as far as possible – to data, source codes and research methods.

The plan is designed to go further than the Law for a Digital Republic, and hinges on three key points: 1) open access to publications; 2) open access to data and the establishment of a sustainable dynamic in France's action; and 3) international activities. An overview of this first plan can be consulted online (MESR 2021a).

A second plan for open science 2021–2024 (MESR 2021b) was published in 2021, and has four main areas of focus: 1) generalizing open access to publications; 2) structuring, sharing and opening up research data; 3) opening up and promoting source codes produced by research; and

1 See: www.ouvrirlascience.fr/home/.

4) transforming practices to make open science the default principle. Practically, the Committee for Open Science is organized into colleges and expert groups around 9 themes: research data; skills and training; publications; software and source codes; assessment; Europe and the international scene; usage analysis; legal and governance. Numerous guides and documents (MESR 2018) have been produced to help the whole community move towards open research.

Scientific publications are more easily accessible thanks to the open archive HAL (n.d.), chosen by the majority of France's scientific and university community. The platform provides long-term preservation of publications and offers numerous services to help unleash their true value, such as the constitution of collections or documentary oversight.

14.1.3. *International context*

At an international level, the Committee CODATA (n.d.) was founded in 1966 to deal specifically with research data. It is part of the International Science Council (n.d.): an NGO formed by the merging of the International Council for Science (founded in 1931) with the International Social Science Council (founded in 1952). This committee has the objective of promoting international collaboration with a view to improving data usage and availability in all scientific domains.

More recently, in 2013, the Research Data Alliance (n.d.b) was formed, in response to the need for open and interoperable sharing of research data, which establishes the social, technical and interdisciplinary links to allow for global sharing (Berman 2019). The consortium is organized into working groups and interest groups, but also "communities of practice" offering a neutral space where members can come together to develop and adopt infrastructures which favor data sharing and data-centered research. Thus, the consortium publishes a number of best-practice guides and recommendations for its community. There are many national RDA groups, including one for France, led by the CNRS, which is in charge of developing the RDA French node. It was initially put in place as part of the project RDA Europe 4.0 (RDA n.d.a). The ministère de l'Enseignement supérieur, de la Recherche et de l'Innovation (MESRI, Ministry for Higher Education, Research and Innovation) supports the operation of the national node, as part of the national plan for open science. This space is dedicated to RDA activities carried out in France and/or by the French community.

14.2. How can we ensure FAIR data in practice?

14.2.1. *Draw inspiration from what exists already*

The computing community was pioneering in engineering openness and the sharing of research data. The concept of free software (Wikipedia n.d.b) emerged in the 1980s, coined by Richard Stallman who, along with others, later formalized and popularized it with the GNU project and the Free Software Foundation (FSF). He specified the concept of free software, and published the GNU general public license (Wikipedia n.d.c), which used the copyright to ensure permanent freedoms awarded to users (and thus precluding the possibility that a later version could no longer be free). The vast majority of free software tends to respect standard open formats, which is conducive to interoperability. In the computing world, the quality of software is often correlated to the number of people having worked on it. The broader the development community becomes, the more the software becomes a gage of quality and responsiveness (Wikipedia n.d.b).

In order to be able to collaborate on the development of a piece of software, we must have access to its source code. The source code is the fundamental building block: it is a text, written in one or more programming languages. To make it accessible, it needs to be made "open", which must be governed by a very specific type of license. The Open Source Movement (Wikipedia n.d.b) applies to software whose license conforms to the criteria set by the Open Source Initiative; namely, the possibility of "free redistribution, source code and derived works" (Open Source Initiative 1998). This movement, which was born in the private sector, relating in particular to software development and distribution methods, developed in parallel to the "free software" movement, the latter having arisen in a university environment, with emphasis placed on the "philosophical and political purpose of the license". There is a whole set of licenses for free/open-source software depending on the types of services they provide, listed on the Open Source Initiative website. In France, the inter-ministry digital directorate (DINUM 2019) is in charge of coordinating actions taken by administrative bodies in relation to information systems. One of its departments, "Etatlab", coordinates the design and implementation of the state's data strategy. The

Web page lists all of the platforms available and many other resources, in particular, a list of existing free licenses[2].

Whilst the domain of information technology is an inspiration in relation to openness of source codes, it can also serve as an inspiration in relation to practices for the development and organization of research. The academic sphere, including all disciplines, should draw more inspiration from these practices and adopt the available development tools. For example, we could point to task automation, code reuse, version management, dependency management and continuous integration. These tasks are part of the life cycle of a piece of software. Very recently, a collective even published an article advocated the use of GitHub[3], a Web platform for software development management, which allows people to collaborate using the Git[4] version control system to speed up collaborative and reproducible research in the fields of ecology and evolution (Braga et al. 2023).

Finally, the FAIR principles (which dictate that data should be Findable, Accessible, Interoperable and Reusable), set forth in 2016 by Wilkinson et al. played a crucial role in the opening up of science. This article marked a turning point in the academic sphere. For the first time, rules and practical recommendations were made available to the scientific community. The FAIR principles are based on the inclusion of metadata in addition to data, in order to make those data more findable, accessible, interoperable and reusable. The metadata must be associated with a global identifier, which is both permanent and, above all, unique. All of these principles are based on the implementation of existing standards specific to the different scientific domains.

14.2.2. *Implementing standards*

What is a standard? In contrast to norms, which are established by a norm-setting organization officially recognized by states, through a standardization organization – e.g. AFNOR (AFNOR n.d.) in France, or the International Standardization Organization (ISO n.d.) – standards are recommendations set by a group of typical users and specialists in a research domain.

2 See: www.data.gouv.fr/fr/pages/legal/licences/.

3 See: www.github.com.

4 See: www.git-scm.com/.

Standards facilitate communication, at a range of different levels. If we look at human interactions, then language (and thus, the commonly spoken language) between human beings could be considered a metaphor for a standard. Similarly, in relation to interacting with a computer, we humans use a set of tools including operating systems and software, through their graphic user interfaces. Finally, computers can also communicate with one another through applications that are specific to them, such as application programming interfaces (APIs) which include standardized tools and protocols facilitating the integration of services and application software.

While standards were first created for the industrial domain to create harmony in a sector's activity, they have been extended to the service domain, notably by norms relating to processes and quality of service.

14.2.3. *Biodiversity Informatics*

Standards ensure, amongst other things, interoperability (the I of FAIR) of data and protocols, by establishing shared frameworks of reference. Natural sciences have been populated with standards specific to the disciplinary field, and the term "biodiversity informatics" (the application of informatic methods and techniques, applied to the study of biodiversity) was coined in 1992[5]. For example, at national level, the *Inventaire national du patrimoine naturel* (INPN n.d.b, National Inventory of Natural Heritage), under the information system of the *Inventaire du patrimoine naturel* (INPN n.d.c) created a standard for the "occurrence of a taxon", which serves to define a nationally shared format (elementary exchange data), compatible with the existing international standards. In the field of biodiversity informatics, an international consortium has been working since 1986 on creating standards for different domains linked to bio- and geodiversity: the Biodiversity Information Standards (TDWG[6]). It is a nonprofit, scientific and educational association whose mission is to create, maintain and promote the use of open standards for biodiversity data to assist with research and decision-making. Its core focus is on the development of a standard used to promote more efficient, wider-scale diffusion and

5 See: www.bgbm.org/BioDivInf/TheTerm.htm.

6 See: www.tdwg.org.

interoperability of such information. For these purposes, the consortium establishes international collaboration between the producers, managers and users of biodiversity information. It develops, ratifies and promotes standards and recommendations on the recording and exchange of data on living and extinct organisms. It serves as a discussion forum on all aspects of information management relating to bio- and geodiversity, through meetings, online discussions and publications. The consortium is organized into an executive board comprising a chair, vice-chair, secretary treasurer, regional representatives for each continent and functional subcommittees, all of the elected members, in addition to the *ex officio* members: a coordinator and an editor in chief for the journal *Biodiversity Information Science and Standards*[7], which publishes the proceedings of the annual conference.

The consortium includes 9 interest groups and 21 working groups (see Figure 14.2), working on topics as varied and complementary as minimum information for a digital specimen (MIDS (Hardisty et al. 2020)), particularly useful for collection specimens, environmental DNA or invasive organisms.

The creation of a standard always begins with one or more ideas that people put forward to the community, and seek to investigate in greater depth (see Figure 14.3). A working group is then set up, under the umbrella of an interest group. The key concepts and associated vocabulary are then compiled in a document which is submitted for ratification (sent for review to at least two reviewers outside of the working group, followed by a public review phase) by the consortium. The functional subcommittee "Technical Architecture" is charged with ensuring interoperability between all of the standards proposed in the consortium's ecosystem. When standards are ratified and adopted, they are monitored by maintenance groups, of which the TDWG currently has four. These groups are responsible for maintaining and updating the existing standards. Each proposed modification goes through a process of public evaluation by the community, and then needs to be ratified by the executive board.

7 See: https://biss.pensoft.net.

Figure 14.2. *Interest groups and working groups of Biodiversity Information Standards (TDWG). For a color version of this figure, see www.iste.co.uk/nicolas/biodiversity.zip*

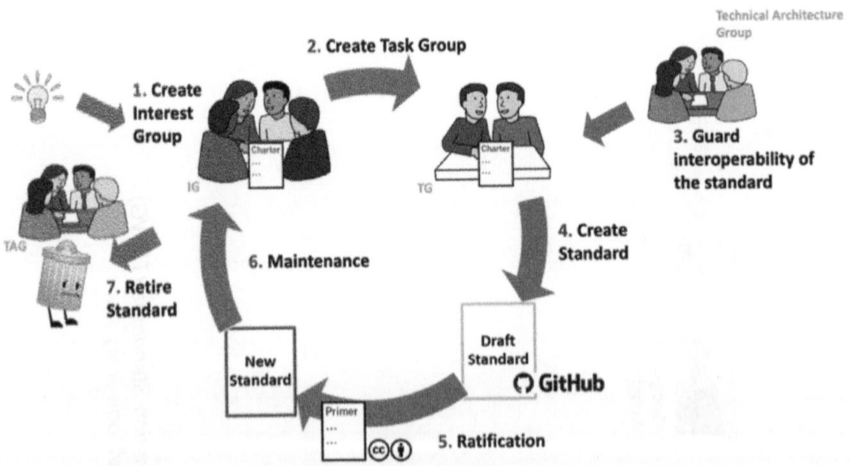

Figure 14.3. *Steps in the creation of a standard. For a color version of this figure, see www.iste.co.uk/nicolas/biodiversity.zip*

Well-known standards include the Darwin Core (Darwin Core Task Group 2009) on occurrence data, which is implemented by the *Global Biodiversity Information Facility*[8], the standard Access to Biological Collection Data (ABCD 2005) and its Extension For Geosciences (Petersen et al. 2018), Audiovisual (formerly Audubon) Core (GBIF/TDWG Multimedia Resources Task Group 2013) on multimedia data, and the Taxonomic Concept Schema (Taxonomic Names and Concepts Interest Group 2005).

TDWG standards have been adopted by many projects and/or initiatives, both European and international. Among these is Europeana[9] – a project offering digital access to European cultural heritage resources – the German Federation for Biological Data[10], and the Global Genome Biodiversity Network[11], which is an international network of institutions that provides a platform for accessing genomic biodiversity databanks.

8 See: www.gbif.org.

9 See: www.europeana.eu/en.

10 See: www.gfbio.org/.

11 See: https://wiki.ggbn.org/ggbn/About_GGBN.

The consortium is also a partner to numerous international and European initiatives, such as the *Alliance for Biodiversity*[12], whose goal is to bring together an alliance around knowledge on biodiversity, and the project *Biodiversity Community Integrated Knowledge Library* (BiCIKL 2020), which is aimed at connecting infrastructures to allow researchers to access services throughout the life cycle of biodiversity data.

Biodiversity Information Standards (TDWG) is an organization grounded in international collaboration, open to one and all. The TDWG is made up of scientists from varied and complementary disciplines, such as informatics, biology and geology, but is not limited to these. Its main strength lies in its broad international community, and all of its productions stem from the expertise, collaboration and support of its members: people and organizations who create, manage, integrate and analyze information relating to biodiversity. It is a key player in the biodiversity data research ecosystem.

14.3. Where can research data be stored? Data warehouses and data papers

In order to make data open and accessible, we need to know which warehouses to store them in and make them available. Every time we store a dataset, it is assigned a unique identifier (generally a DOI) to make it more easily findable and accessible (the F and A in FAIR).

At national level, there are multiple possibilities open to academics. The MESRI has adopted a general and global ecosystem (covering all domains and disciplines): Recherche Data Gouv[13], and offers guidance through its platform launched in 2022. This open, community platform is aimed at centralizing and structuring open data in France. Three modules are designed to guide research teams, offering data workshops, topic-specific reference centers and resource hubs. The platform presents two modules directly linked to the storing, publication and referencing of data: it has a data warehouse for the storage and use of data, and a catalog to search data published in both this and external warehouses. Among the topic-specific reference centers is the *Pôle national de données de biodiversité*[14] (National Biodiversity Data Hub): a virtual research infrastructure offering tools and

12 See: www.allianceforbio.org/.

13 See: https://recherche.data.gouv.fr/en.

14 See: www.pndb.fr/en/.

services for sharing and use of both data and metadata, specifically relating to biodiversity data. The project *Inventaire des données de la recherche en environnement et sociétés*[15] (Environmental and Societal Research Data Inventory) – part of the research and support unit *Bases de données Biodiversité, Écologie, Environnement et Sociétés* (BBEES[16], Biodiversity, Ecology, Environment and Society Databases), run jointly by the CNRS and the MNHN – provides a platform also offering tools and services relating to biodiversity data and the storage of such data. The InDoRES portal has a standardized and interoperable catalog of metadata which can be used to browse existing databases and datasets[17], and a secure data warehouse[18].

It is also possible to use the tools and services offered by the OPIDoR[19] platform, run by the CNRS' *Institut de l'information scientifique et technique* (Institute for Scientific and Technical Information). One such tool is a support to create a data-management plan (DMP OPIDoR), a wiki catalog of services dedicated to research data (Cat OPIDoR) and a service for assigning persistent identifiers (PID OPIDoR). This is a multidisciplinary platform, unlike the previous two.

At a European scale, the European Open Science Cloud[20] project is aimed at creating a virtual environment for hosting, sharing and access to European research data. This multidisciplinary platform is distinguished by the fact that it offers a whole range of complementary services, including access to supercomputers, analysis tools and training. The college EOSC-France represents the interests of French scientific communities.

Internationally, the GBIF network is certainly the best known. It is an international network, and also a data infrastructure, funded by the member states to provide free and open access to biodiversity data. It is possible to "deposit" a dataset on the GBIF portal, using the Darwin Core standard format published by the *Biodiversity Information Standards* (TDWG) mentioned above.

15 See: www.indores.fr/.

16 See: www.bbees.mnhn.fr/fr.

17 See: www.indores.fr/geonetwork/srv/fre/catalog.search#/home.

18 See: https://data.indores.fr/.

19 See: www.opidor.fr/.

20 See: www.eosc-portal.eu/.

It is also possible to accompany the publication of a dataset in an open repository by publishing a "data paper", possibly but not necessarily beforehand. These scientific articles have the peculiarity of containing only the raw data and the associated metadata to describe those data. Various so-called data journals specialize in this type of publication, pertaining to various scientific disciplines. The Cirad portal[21] presents a list of journals in which it is possible to publish data papers. The IRD also has a similar platform[22]. The main contribution of a data paper is that it allows the reuse (*the R of FAIR*) of the published data. Thus, these datasets must comply with disciplinary standards recognized by the various scientific communities and implement best practices. Doing this allows the data to be valorized, facilitates their reuse, gives them greater visibility and, finally, makes them more accessible and findable (they can be cited directly, because they are in scientific publications just like a "conventional" scientific publication). Data papers are also reviewed by external reviewers before being accepted for publication.

14.4. Conclusion

14.4.1. *From the need to shift the sociocultural paradigm...*

Globalization is also being seen in the academic and scientific worlds. Technological advances have made for huge progress in many, many areas of science, for example, structural or functional genomics, bio-informatics and the development of artificial intelligence, which has become a catalyst for more old-fashioned disciplines such as taxonomy. Indeed, photographic identification systems, used in combination with learning algorithms, will soon become a tool in scientists' kit (Shirai et al. 2022). Other tools such as barcoding (Hebert et al. 2003) have recently been developed.

However, what we are witnessing is an unprecedented paradox. We have entered a period described as the sixth mass extinction event (Kolbert 2014; Ceballos et al. 2015, 2020; Régnier et al. 2015; Cowie et al. 2022; and others). Biodiversity has been in decline for many years, as a result of human activities (Myers et al. 2000; Sachs et al. 2009; Butchard et al. 2010) and yet, our technology has never been so advanced. We are presented with a deluge

21 See: https://ou-publier.cirad.fr/revues?f%5B0%5D=themes%3A69.

22 See: https://data.ird.fr/datapapers/.

of results and data which need to be managed and exploited as best they can, to produce benefits for the planet as a whole.

The aim here is to bring about a change that could be described as a sociocultural paradigm shift. We need to produce a change in the culture of data science and research science. It is imperative that we engage in building a community for open research and open science, build bridges between the different disciplines and promote interdisciplinarity. Open science is a new way of conducting research, bringing together the production and dissemination of research, so that as many people as possible can access it freely (in both senses of the word), reuse it, build and distribute these resources, without hampering it.

14.4.2. *...to more ethical research*

Here, the idea is to integrate an ethical dimension into research, which is what open science encourages. Making research data accessible to all promotes the idea of inclusion. In 2010, the Convention on Biological Diversity[23] adopted the "Nagoya Protocol on Access to Genetic Resources and Sharing of Benefits Arising from their use", which came into force in October 2014. This protocol marked the first step towards more ethical research. More recently, in 2020, under the aegis of the RDA (n.d.b), the "Global Indigenous Data Alliance" (GIDA[24]) was set up, with the goal of encouraging Indigenous people's control of indigenous data. This alliance is made up of Indigenous researchers, users and activists who defend the sovereignty and governance of indigenous data within their Nation State and internationally. In 2020, the members of GIDA published the CARE principles (Collective Benefit, Authority to control, Responsibility, Ethics) to go along with the FAIR principles (Caroll et al. 2020, 2022a, 2022b) (see Figure 14.4), which emphasize taking account of the rights and sovereignty of indigenous peoples when the world is engaging in open science and data sharing. In the same spirit of inclusion and fairness, scientists have proposed "10 simple rules to conduct fairer field missions in biology" (Ramirez-Castaneda et al. 2022).

23 See: www.cbd.int/cop10.

24 See: www.gida-global.org/.

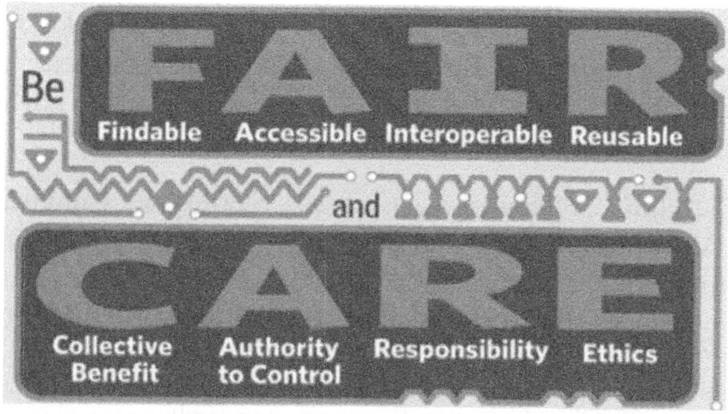

Figure 14.4. *Be FAIR and CARE (Caroll et al. 2020, 2022a, 2022b). For a color version of this figure, see www.iste.co.uk/nicolas/biodiversity.zip*

14.5. References

AFNOR (n.d.). La normalisation [Online]. Available at: https://normalisation.afnor.org/ [Accessed 31 March 2024].

Berlin Declaration (2003). Berlin declaration on open access to knowledge in the sciences and humanities [Online]. Available at: https://openaccess.mpg.de/Berlin-Declaration [Accessed 31 March 2024].

Berman, F. (2019). The Research Data Alliance – The first five years. Document, RDA, Oxford.

Berman, F. and Crosas, M. (2020). The research data alliance: Benefits and challenges of building a community organization. *Harvard Data Science Review*, 2(1). doi: 10.1162/99608f92.5e126552.

BiCIKL (2020). Biodiversity community integrated knowledge library [Online]. Available at: https://bicikl-project.eu/ [Accessed 31 March 2024].

Biodiversity Information Standards (TDWG) (ed.) (2005). Access to biological collection data (ABCD) [Online]. Available at: http://www.tdwg.org/standards/115 [Accessed 31 March 2024].

Braga, P.H.P., Hébert, K., Hudgins, E.J., Scott, E.R, Edwards, B.P.M., Sánchez Reyes, L.L., Grainger, M.J., Foroughirad, V., Hillemann, F., Binley, A.D. et al. (2023). Not just for programmers: How GitHub can accelerate collaborative and reproducible research in ecology and evolution. *Methods in Ecology and Evolution*, 14(6), 1364–1380. doi: 10.1111/2041-210X.14108.

Butchart, S.H.M., Walpole, M., Collen, B., Van Strien, A., Scharlemann J.P.W., Almond, R.E.A., Baillie, J.E.M., Bomhard, B., Brown, C., Bruno, J. et al. (2010). Global biodiversity: Indicators of recent declines. *Science*, 328, 1164–1168.

Carroll, S.R., Garba, I., Figueroa-Rodríguez, O.L., Holbrook, J., Lovett, R., Materechera, S., Parsons, M., Raseroka, K., Rodriguez-Lonebear, D., Roweet, R. et al. (2020). The CARE principles for indigenous data governance. *Data Science Journal*, 19(43), 1–12. doi: 10.5334/dsj-2020-043.

Carroll, S.R., Plevel, R., Jennings, L.L., Garba, I., Sterling, R., Cordova-Marks, F.M., Hiratsuka, V., Hudson, M., Garrison, N.A. (2022a). Extending the CARE principles from tribal research policies to benefit sharing in genomic research. *Frontiers in Genetics*, 13(1052620). doi: 10.3389/fgene.2022.1052620.

Carroll, S.R., Garba, I., Plevel, R., Small-Rodriguez, D., Hiratsuka, V.Y., Hudson, M., Garrison, N.A. (2022b). Using indigenous standards to implement the CARE principles: Setting expectations through tribal research codes. *Frontiers in Genetics*, 13(823309). doi: 10.3389/fgene.2022.823309.

Ceballos, G., Ehrlich, P.R., Barnosky, A., García, A., Pringle, R., Palmer, T. (2015). Accelerated modern human-induced species losses: Entering the sixth mass extinction. *Science Advances*, 1, e1400253.

Ceballos, G., Ehrlich, P.R., Raven, P.H. (2020). Vertebrates on the brink as indicators of biological annihilation and the sixth mass extinction. *Proceedings of the National Academy of Sciences USA*, 117, 13596–13602.

CODATA (n.d.). About Codata [Online]. Available at: https://codata.org/about-codata/ [Accessed 5 June 2023].

Cowie, R.H., Bouchet, P., Fontaine, B. (2022). The sixth mass extinction: Fact, fiction or speculation? *Biological Review*, 97(2), 640–663.

Darwin Core Task Group (2009). Darwin core. Biodiversity Information Standards (TDWG) [Online]. Available at: http://www.tdwg.org/standards/450 [Accessed 31 March 2024].

Direction interministérielle du numérique (2019). La DIMUM. Le numérique au service de l’efficacité de l’action publique [Online]. Available at: https://www.numerique.gouv.fr/dinum/ [Accessed 31 March 2024].

GBIF/TDWG Multimedia Resources Task Group (2013). Audiovisual core multimedia resources metadata schema. Biodiversity Information Standards (TDWG) [Online]. Available at: http://www.tdwg.org/standards/638 [Accessed 31 March 2024].

Gomez-Diaz, T. (2023). Les logiciels de la recherche et leurs licences : trois visions sur un objet. Master’s Thesis, Université Gustave Eiffel, Marne-la-Vallée.

HAL Science ouverte (n.d.). HAL [Online]. Available at: https://hal.science/ [Accessed 5 June 2023].

Hardisty, A.R., Addink, W., Dillen, M., Groom, Q., Haston, E., Glöckler, F., Paul, D., Petersen, M., Saarenmaa, H., Güntsch, A. (2020). Minimum information about a digital specimen (MIDS). *Draft Text of a Specification*, 11.

Hebert, P.D.N., Cywinska, A., Ball, S.L., deWaard, J.R. (2003). Biological identifications through DNA barcodes. *Proceedings of the Royal Society B: Biological Sciences*, 270313–270321. doi: 10.1098/rspb.2002.2218.

iNaturalist (n.d.). iNaturalist [Online]. Available at: https://www.inaturalist.org [Accessed 31 March 2024].

INPN (n.d.a). INPN [Online]. Available at: https://inpn.mnhn.fr/accueil/participer/inpn-especes?lg=en [Accessed 31 March 2024].

INPN (n.d.b). INPN [Online]. Available at: https://inpn.mnhn.fr/accueil/index [Accessed 31 March 2024].

INPN (n.d.c). Le SINP, système d'information de l'inventaire du patrimoine [Online]. Available at: https://inpn.mnhn.fr/informations/sinp/presentation [Accessed 31 March 2024].

International Organisation for Standardization (n.d.). ISO [Online]. Available at: www.iso.org/home.html [Accessed 31 March 2024].

International Science Council (n.d.). [Online]. Available at: https://council.science/ [Accessed 31 March 2024].

Inventaire des données de la recherche en environnement et sociétés (n.d.). Indores [Online]. Available at: www.indores.fr/ [Accessed 31 March 2024].

Kolbert, E. (2014). *The Sixth Extinction. An Unnatural History*. Henry Holt and Company, New York.

Légifrance (2016). Loi no. 2016-1321 du 7 octobre 2016 pour une République numérique (1) [Online]. Available at: https://www.legifrance.gouv.fr/jorf/article_jo/JORFARTI000033203472 [Accessed 31 March 2024].

Les Herbonautes (n.d.). Bibliothèque de la science ouverte [Online]. Available at: https://www.ouvrirlascience.fr/category/resources/ [Accessed 31 March 2024].

MESR (2018). Plan national pour la science ouverte [Online]. Available at: https://www-test.ouvrirlascience.fr/wp-content/uploads/2018/08/PLAN_NATIONAL_SCIENCE_OUVERTE_978672.pdf [Accessed 21 May 2024]

MESR (2021a). Bilan du Plan national pour la science ouverte 2018–2021 [Online]. Available at: https://www.ouvrirlascience.fr/bilan-du-plan-national-pour-la-science-ouverte-2018-2021/ [Accessed 31 March 2024].

MESR (2021b). Deuxième Plan national pour la science ouverte. Généraliser la science ouverte en France 2021–2024 [Online]. Available at: https://www.enseignementsup-recherche.gouv.fr/sites/default/files/content_migration/document/2e-plan-national-pour-la-science-ouverte-2021-2024-7794.pdf [Accessed 31 March 2024].

Myers, N., Mittermeier, R.A., Mittermeier, C.G., da Fonseca, G.A., Kent, J. (2000). Biodiversity hotspots for conservation priorities. *Nature*, 403(6772), 853–858.

OECD (2007). OECD principles and guidelines for access to research data from public funding [Online]. Available at: https://www.oecd-ilibrary.org/docserver/9789264034020-en-fr.pdf [Accessed 31 March 2024].

Open Source Initiative (1998). [Online]. Available at: https://opensource.org/ [Accessed 31 March 2024].

Petersen, M., Glöckler, F., Kiessling, W., Döring, M., Fichtmüller, D., Laphakorn, L., Baltruschat, B., Hoffmann, J. (2018). History and development of ABCDEFG: A data standard for geosciences. *Fossil Record*, 21(1), 47–53. doi: 10.5194/fr-21-47-2018.

PNDB (n.d.). Pôle national de données de biodiversité [Online]. Available at: https://www.pndb.fr/fr/ [Accessed 31 March 2024].

Ramirez-Castaneda, V., Westeen, E.P., Frederick, J., Amini, S., Wait, D.R., Achmadi, A.S., Andayani, N., Arida, E., Arifin, U., Bernal, M.A. et al. (2022). A set of principles and practical suggestions for equitable field works in biology. *PNAS*, 119(34), e2122667119.

RDA (n.d.a). About RDA Europe 4.0 [Online]. Available at: https://grants.rd-alliance.org/about-rda-eu-40 [Accessed 31 March 2024].

RDA (n.d.b). Research Data Alliance [Online]. Available at: https://www.rd-alliance.org/ [Accessed 31 March 2024].

Recherche Data Gouv (n.d.). An ecosystem for sharing and opening research data [Online]. Available at: https://recherche.data.gouv.fr/en [Accessed 31 March 2024].

Sachs, J.D., Baillie, J.E.M., Sutherland, W.M., Armsworth, P.R., Ash, N., Beddington, J., Blackburn T.M., Jones K.E. (2009). Biodiversity conservation and the millennium development goals. *Science*, 325, 1502–1503.

Shirai, M., Takano, A., Kurosawa, T., Inoue, M., Tagane, S., Tanimoto, T., Koganeyama, T., Sato, H., Terasawa, T., Horie, T. et al. (2022). Development of a system for the automated identification of herbarium specimens with high accuracy. *Sci Rep.*, 12, 8066. doi: 10.1038/s41598-022-11450-y.

Taxonomic Names and Concepts Interest Group. (2005). Taxonomic concept transfer schema (TCS). Biodiversity Information Standards (TDWG) [Online]. Available at: http://www.tdwg.org/standards/117 [Accessed 31 March 2024].

The Turing Way Community (2021). The Turing Way: A handbook for reproducible, ethical and collaborative research (1.0.1). *Zenodo*. doi: 10.5281/zenodo.5671094.

Wikipedia (n.d.a). Open data [Online]. Available at: https://en.wikipedia.org/wiki/Open_data [Accessed 31 March 2024].

Wikipedia (n.d.b). Free software [Online]. Available at: https://en.wikipedia.org/wiki/Free_software [Accessed 31 March 2024].

Wikipedia (n.d.c). GNU general public license [Online]. Available at: https://en.wikipedia.org/wiki/GNU_General_Public_License [Accessed 31 March 2024].

Wikipedia (n.d.d). Open source [Online]. Available at: https://en.wikipedia.org/wiki/Open_source [Accessed 31 March 2024].

Wilkinson, M., Dumontier, M., Aalbersberg, I., Appleton, G., Axton, M., Baak, A., Blomberg, N., Boiten, J.-W., Bonino da Silva Santos, L.O., Bourne, P.E. et al. (2016). The FAIR guiding principles for scientific data management and stewardship. *Scientific Data*, 3(160018). doi: 10.1038/sdata.2016.18.

List of Authors

Guillaume ACHAZ
CIRB
Université Paris-Cité
France

Nadia AMÉZIANE
ISYEB, CNRS
Muséum national d'Histoire naturelle
Sorbonne Université
EPHE-PSL
Université des Antilles
France

Cécile BANOVSKI-KLEIN
LEMAR
Université de Bretagne Occidentale
Plouzané
France

Sophie BARY
ISYEB, CNRS
Muséum national d'Histoire naturelle
Sorbonne Université
EPHE-PSL
Université des Antilles
France

Marta BELLATO
2ADN, CNRS
Muséum national d'Histoire naturelle
Paris
France

Julien BIGOT
ISYEB, CNRS
Muséum national d'Histoire naturelle
Sorbonne Université
EPHE-PSL
Université des Antilles
France

Mathieu BODERAU
ISYEB, CNRS
Muséum national d'Histoire naturelle
Sorbonne Université
EPHE-PSL
Université des Antilles
France

Frédéric BUSSON
BOREA, CNRS
IRD
Muséum national d'Histoire naturelle
Sorbonne Université
Université de Caen Normandie
Université des Antilles
France

Mathilde CADIER
Laboratoire d'écologie pélagique
IFREMER, DYNECO, PELAGOS
Plouzané
France

Romain CAUSSE
BOREA, CNRS
IRD
Muséum national d'Histoire naturelle
Sorbonne Université
Université de Caen Normandie
Université des Antilles
France

Antoine CHARPENTIER
Observatoire du Plancton
Port-Louis
France

Gaël DENYS
PatriNat
Office français de la biodiversité
Muséum national d'Histoire naturelle
Paris
France

Agnès DETTAÏ
ISYEB, CNRS
Muséum national d'Histoire naturelle
Sorbonne Université
EPHE-PSL
Université des Antilles
France

Marc ELÉAUME
ISYEB, CNRS
Muséum national d'Histoire naturelle
Sorbonne Université
EPHE-PSL
Université des Antilles
France

Nathalie ELISABETH
LRC Systems
Menlo Park
and
Lawrence Berkeley National Laboratory
United States

Klervi FUSTEC
AMURE, IUEM
Université de Bretagne Occidentale
Plouzané
France

Cyril GALLUT
ISYEB, CNRS
Muséum national d'Histoire naturelle
Sorbonne Université
EPHE-PSL
Université des Antilles
France

Romain GARROUSTE
ISYEB, CNRS
Muséum national d'Histoire naturelle
Sorbonne Université
EPHE-PSL
Université des Antilles
France

Silvina GONZALEZ-RIZZO
ISYEB
Université des Antilles
Pointe-à-Pitre
Guadeloupe

Olivier GROS
ISYEB
Université des Antilles
Pointe-à-Pitre
Guadeloupe

Corentin JOUAULT
ISYEB, CNRS
Muséum national d'Histoire naturelle
Sorbonne Université
EPHE-PSL
Université des Antilles
France

Jean-Luc JUNG
ISYEB, CNRS
Muséum national d'Histoire naturelle
Sorbonne Université
EPHE-PSL
Université des Antilles
France

Philippe KEITH
BOREA, CNRS
IRD
Muséum national d'Histoire naturelle
Sorbonne Université
Université de Caen Normandie
Université des Antilles
France

Anna KONDRATYEVA
ISYEB, CNRS
Muséum national d'Histoire naturelle
Sorbonne Université
EPHE-PSL
Université des Antilles
France

Jean-Paul KUNDURA
ISYEB, CNRS
Muséum national d'Histoire naturelle
Sorbonne Université
EPHE-PSL
Université des Antilles
France

Line LE GALL
ISYEB, CNRS
Muséum national d'Histoire naturelle
Sorbonne Université
EPHE-PSL
Université des Antilles
France

Julianne LE GUEN
Océanopolis
Brest
France

Marine LE MOAL
Océanopolis
Brest
France

Molène LE ROY
Explore
Concarneau
France

Guillaume LECOINTRE
ISYEB, CNRS
Muséum national d'Histoire naturelle
Sorbonne Université
EPHE-PSL
Université des Antilles
France

Antoine LÉVÊQUE
PatriNat
Office français de la biodiversité
Muséum national d'Histoire naturelle
Paris
France

Céline LIRET
Océanopolis
Brest
France

Pablo MARTINEZ-SOARES
ISYEB, CNRS
Muséum national d'Histoire naturelle
Sorbonne Université
EPHE-PSL
Université des Antilles
France

Margot MATHIEU
ISYEB, CNRS
Muséum national d'Histoire naturelle
Sorbonne Université
EPHE-PSL
Université des Antilles
France

Aurélien MIRALLES
Zoological Institute
Technical University of Braunschweig
Germany

Romain NATTIER
ISYEB, CNRS
Muséum national d'Histoire naturelle
Sorbonne Université
EPHE-PSL
Université des Antilles
France

André NEL
ISYEB, CNRS
Muséum national d'Histoire naturelle
Sorbonne Université
EPHE-PSL
Université des Antilles
France

Patricia NEL
ISYEB, CNRS
Muséum national d'Histoire naturelle
Sorbonne Université
EPHE-PSL
Université des Antilles
France

Valérie NGÔ-MULLER
ISYEB, CNRS
Muséum national d'Histoire naturelle
Sorbonne Université
EPHE-PSL
Université des Antilles
France

Violaine NICOLAS
ISYEB, CNRS
Muséum national d'Histoire naturelle
Sorbonne Université
EPHE-PSL
Université des Antilles
France

Catherine OZOUF-COSTAZ
ISYEB, CNRS
Muséum national d'Histoire naturelle
Sorbonne Université
EPHE-PSL
Université des Antilles
France

Henri PERSAT
SFI
Paris
France

Joëlle PICHON
Océanopolis
Brest
France

Martin PLUS
Laboratoire d'écologie pélagique
Ifremer, DYNECO, PELAGOS
Plouzané
France

Philippe PONDAVEN
LEMAR-IUEM
Université de Bretagne Occidentale
Plouzané
France

Nicolas POULET
OFB
Toulouse
France

Nicolas PUILLANDRE
ISYEB, CNRS
Muséum national d'Histoire naturelle
Sorbonne Université
EPHE-PSL
Université des Antilles
France

Rodolphe ROUGERIE
ISYEB, CNRS
Muséum national d'Histoire naturelle
Sorbonne Université
EPHE-PSL
Université des Antilles
France

Michel ROUX
ISYEB, CNRS
Muséum national d'Histoire naturelle
Sorbonne Université
EPHE-PSL
Université des Antilles
France

Karen SALAZAR
ISYEB, CNRS
Muséum national d'Histoire naturelle
Sorbonne Université
EPHE-PSL
Université des Antilles
France

Sarah SAMADI
ISYEB, CNRS
Muséum national d'Histoire naturelle
Sorbonne Université
EPHE-PSL
Université des Antilles
France

Nalani K. SCHNELL
ISYEB, CNRS
Muséum national d'Histoire naturelle
Sorbonne Université
EPHE-PSL
Université des Antilles
France

Thomas SCHUBNEL
ISYEB, CNRS
Muséum national d'Histoire naturelle
Sorbonne Université
EPHE-PSL
Université des Antilles
France

Laura SCHWEIBOLD
LEMAR-IUEM
Université de Bretagne Occidentale
Plouzané
France

Lucas SIRE
ISYEB, CNRS
Muséum national d'Histoire naturelle
Sorbonne Université
EPHE-PSL
Université des Antilles
France

Marc SOURISSEAU
Laboratoire d'écologie pélagique
IFREMER, DYNECO, PELAGOS
Plouzané
France

Visotheary UNG
ISYEB, CNRS
Muséum national d'Histoire naturelle
Sorbonne Université
EPHE-PSL
Université des Antilles
France

Jean-Marie VOLLAND
LRC Systems
Menlo Park
and
Lawrence Berkeley National
Laboratory
United States

Patricia WILLS
2AD, CNRS
Muséum national d'Histoire naturelle
Paris
France

Index

Other titles from

iSTE

in

Biology and Biomedical Engineering

2024

MASQUIDA Benoît, LECLERC Fabrice
Looking at Ribozymes: Biology of Catalytic RNA
(Nucleic Acids Set – Volume 2)

2023

BALTHAZART Jacques
Fifty Years of Evolution in Biological Research: Progress and Decline

GILBERT Marc, PIRKMAJER Sergej
Concepts in Biology: A Historical Perspective

VERNA Emeline
Asymptomatic Osseous Variations of the Postcranial Human Skeleton
(Comparative Anatomy and Posture of Animal and Human Set – Volume 5)

WALCH Jean-Paul, BLAISE Solange
Phyllotaxis Models: A Tool for Evolutionary Biologists

2022

DAMBRICOURT MALASSÉ Anne
Embryogeny and Phylogeny of the Human Posture 2: A New Glance at the Future of our Species
(Comparative Anatomy and Posture of Animal and Human Set – Volume 4)

FOSSÉ Philippe
Structures and Functions of Retroviral RNAs: The Multiple Facets of the Retroviral Genome
(Nucleic Acids Set – Volume 1)

2021

DAMBRICOURT MALASSÉ Anne
Embryogeny and Phylogeny of the Human Posture 1: A New Glance at the Future of our Species
(Comparative Anatomy and Posture of Animal and Human Set – Volume 3)

GRANDCOLAS Philippe, MAUREL Marie-Christine
Systematics and the Exploration of Life

HADJOUIS Djillali
The Skull of Quadruped and Bipedal Vertebrates: Variations, Abnormalities and Joint Pathologies
(Comparative Anatomy and Posture of Animal and Human Set – Volume 2)

HULLÉ Maurice, VERNON Philippe
The Terrestrial Macroinvertebrates of the Sub-Antarctic Îles Kerguelen and Île de la Possession

2020

CAZEAU Cyrille
Foot Surgery Viewed Through the Prism of Comparative Anatomy: From Normal to Useful
(Comparative Anatomy and Posture of Animal and Human Set – Volume 1)

DUJON Bernard, PELLETIER Georges
Trajectories of Genetics

2019

BRAND Gérard
Discovering Odors

BUIS Roger
Biology and Mathematics: History and Challenges

2016

CLERC Maureen, BOUGRAIN Laurent, LOTTE Fabien
Brain–Computer Interfaces 1: Foundations and Methods
Brain–Computer Interfaces 2: Technology and Applications

FURGER Christophe
Live Cell Assays: From Research to Health and Regulatory Applications

2015

CLARYSSE Patrick, FRIBOULET Denis
Multi-modality Cardiac Imaging: Processing and Analysis

2014

CHÈZE Laurence
Kinematic Analysis of Human Movement

DAO Tien Tuan, HO BA THO Marie-Christine
Biomechanics of the Musculoskeletal System: Modeling of Data Uncertainty and Knowledge

FANET Hervé
Medical Imaging Based on Magnetic Fields and Ultrasounds

FARIÑAS DEL CERRO Luis, INOUE Katsumi
Logical Modeling of Biological Systems

MIGONNEY Véronique
Biomaterials

TEBBANI Sihem, LOPES Filipa, FILALI Rayen, DUMUR Didier,
PAREAU Dominique
CO_2 Biofixation by Microalgae: Modeling, Estimation and Control